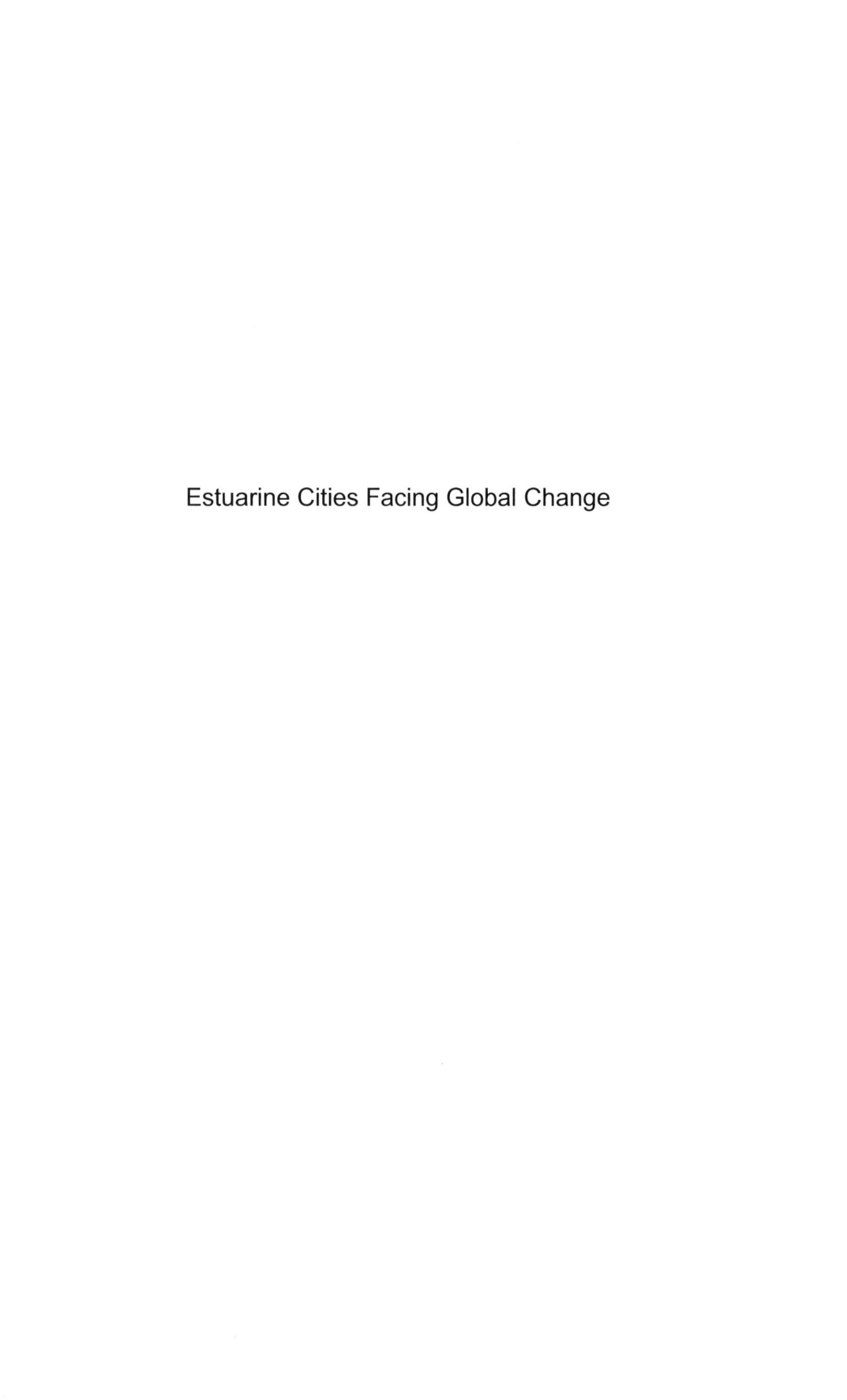

Estuarine Cities Facing Global Change

Series Editor
Françoise Gaill

Estuarine Cities Facing Global Change

Towards Anticipatory Governance

Edited by
Denis Salles
Glenn Mainguy
Charles de Godoy Leski

WILEY

*Published with the support of the Centre Emile Durkheim
(CNRS/Sciences Po Bordeaux/University of Bordeaux/Idex)*

First published 2023 in Great Britain and the United States by ISTE Ltd and John Wiley & Sons, Inc.

Apart from any fair dealing for the purposes of research or private study, or criticism or review, as permitted under the Copyright, Designs and Patents Act 1988, this publication may only be reproduced, stored or transmitted, in any form or by any means, with the prior permission in writing of the publishers, or in the case of reprographic reproduction in accordance with the terms and licenses issued by the CLA. Enquiries concerning reproduction outside these terms should be sent to the publishers at the undermentioned address:

ISTE Ltd
27-37 St George's Road
London SW19 4EU
UK

www.iste.co.uk

John Wiley & Sons, Inc.
111 River Street
Hoboken, NJ 07030
USA

www.wiley.com

© ISTE Ltd 2023

The rights of Denis Salles, Glenn Mainguy and Charles de Godoy Leski to be identified as the authors of this work have been asserted by them in accordance with the Copyright, Designs and Patents Act 1988.

Any opinions, findings, and conclusions or recommendations expressed in this material are those of the author(s), contributor(s) or editor(s) and do not necessarily reflect the views of ISTE Group.

Library of Congress Control Number: 2023930943

British Library Cataloguing-in-Publication Data
A CIP record for this book is available from the British Library
ISBN 978-1-78630-710-1

Contents

About the Authors xi

Acknowledgments xv

Introduction xvii
Denis SALLES, Glenn MAINGUY and Charles DE GODOY LESKI

Part 1. The Water of the Cities and the Water of the Fields 1

Chapter 1. The Governance of Socio-Ecological Interdependencies: The Landes du Médoc Water Catchment Area Controversy 3
Charles DE GODOY LESKI

    1.1. Introduction 3
    1.2. Drinking water supply in Gironde, the history of a transfer from surface water to deep groundwater 6
        1.2.1. Under the Roman Empire: the administration of a city's water sources in its estuary 6
        1.2.2. The Middle Ages: the era of hydraulic and defensive withdrawal of the city 7
        1.2.3. The hydraulic "Renaissance" in the 17th century: towards the golden age of Bordeaux fountains 8
        1.2.4. From 1800 to 1850: the dark age of resource availability 9
        1.2.5. From 1850 to 1950: towards extractivist geopolitics of water 10
        1.2.6. From 1950 to the present day: the socio-technical anticipation of issues through the governance of the drinking water resource 11

1.3. Metropolitan territorial conceptions in the face of forestry
references. 13
1.4. The refinement of models and the rising criticism 19
1.5. Conclusion 23
1.6. References 25

**Chapter 2. Ecological Engineering in a Controversial Drinking
Water Production Project.** 27
Alain DUPUY and Aude VINCENT

2.1. The socio-hydrogeological configuration of the Landes du
Médoc catchment area (Gironde). 28
2.2. An ecological engineering solution. 30
2.3. How much of the extracted water must be reinjected? 32
    2.3.1. Percentage efficiency of the water reinjected 33
    2.3.2. Efficiency of the reinjection duration 34
    2.3.3. Efficiency of the selected method to reinject the water. 34
2.4. When and where should the extracted water be reinjected? 37
2.5. Conclusion 39
2.6. References 39

**Part 2. Protecting Against Risks, by the Estuary, and for
the Estuary.** 43

**Chapter 3. Living in a City Exposed to Flood Risk: At What
Cost(s)?.** 45
Jeanne DACHARY-BERNARD and Florian VERGNEAU

3.1. Residential location and risk as economic issues. 47
3.2. Empirical strategy of the hedonic price model 49
3.3. Bordeaux Métropole study area and data 50
3.4. A multifaceted city 55
    3.4.1. Confirmed metropolitan trends coupled with emerging
    rurbanization 57
    3.4.2. A double effect of flood risk on prices 61
3.5. Conclusion 63
3.6. References 64

**Chapter 4. The Ecological Restoration of Estuaries: Protection
of People and Combating the Erosion of Biodiversity.** 69
Mario LEPAGE, Michael ELLIOTT, Cécile CAPDERREY and
Henrique CABRAL

4.1. Habitats, biodiversity and ecosystem services 71
4.2. Causes of the ecological degradation of estuaries 72

4.2.1. Effects of rising sea levels.	72
4.2.2. Effects of anthropogenic pressures on biodiversity	73
4.3. Ecological restoration of estuaries for the protection of biodiversity.	76
4.3.1. Active and passive restoration	79
4.4. Examples of ecological restoration in estuaries.	81
4.4.1. The marshes of Mortagne-sur-Gironde (France)	81
4.4.2. Mondego estuary (Portugal)	83
4.4.3. Scheldt estuary (Belgium).	85
4.5. Conclusion	88
4.6. References	89

Chapter 5. Sensemaking in the Face of Estuarine Flood Risk Mitigation 93
Jean-Paul VANDERLINDEN and Nabil TOUILI

5.1. The conceptual framework of narrative analysis	94
5.1.1. Stories of risk governance.	95
5.1.2. Sensemaking as a source of narratives about change	96
5.1.3. A corpus of interviews on the risk of flooding in Gironde.	96
5.2. Ethical theories invoked and associated meta-narratives	98
5.2.1. Deontology in terms of having respect for shared norms: the meta-narrative of deontological hype.	98
5.2.2. Virtue ethics: the meta-narrative of "respect for justice" as a virtue.	100
5.2.3. Consequentialism in risk reduction: the meta-narrative of ordinary risk governance	101
5.2.4. Consequentialism in terms of inequity: the meta-narrative of the questionable fairness of choices made	102
5.2.5. Deliberation ethics: the meta-narrative of the process that is to be improved	103
5.2.6. Ethics of nature: the meta-narrative of nature holds the keys	105
5.3. For deliberative risk governance	106
5.4. Conclusion	108
5.5. References	109

Part 3. When the Estuary Makes the City 113

Chapter 6. The Estuarine City as an Allegory for Changes in Solidarity 115
Thierry OBLET

6.1. Cleansing the metropolitan idea of the stench of its emissions and ecological irresponsibility	117

6.2. From the conquest of land to the recognition of territories 121
6.3. From equality to territorial cohesion 124
6.4. Conclusion 130
6.5. References 131

Chapter 7. Nantes and the Loire: Construction of an Estuarine City Faced with Port and Environment Challenges 135
Glenn MAINGUY

7.1. Emergence of the estuarine dimension: from the city of
Nantes–Saint-Nazaire to the opening of the "Terre d'Estuaire"
museum. 137
   7.1.1. From an institutional dimension 137
   7.1.2. … to a cultural and tourist vocation 139
7.2. When Nantes and the Loire drifted apart: a progressive denial of
the city's estuarine dimension. 140
   7.2.1. From digging a canal to filling in waterways 140
   7.2.2. Departure from shipyards 143
7.3. Building a new relationship between Nantes Métropole and its
estuary: the desire for the Loire. 143
   7.3.1. Integration of the estuarine dimension through heritage and
industrial-port memory 143
   7.3.2. The Great Debate: a participatory tool for reclaiming the Loire . . 145
   7.3.3. *Conférence Permanente Loire* and *Mission Loire*:
putting environmental issues related to the Loire on the agenda 147
7.4. Conclusion 149
7.5. References 150

Part 4. Anticipating the Future of Estuarine Cities 153

**Chapter 8. Past and Future Socio-Ecological Pathways of the
Seine Estuary** 155
Gilles BILLEN, Julia LE NOË, Camille NOÛS and Josette GARNIER

8.1. The Seine estuary as a socio-ecological system 156
8.2. The successive phases of port traffic. 159
8.3. The energy supply of the Seine basin 160
8.4. The contribution of ports to the agri-food system of the Seine
basin. 162
8.5. The era of globalized trade in manufactured goods 165
8.6. What is the future of the Seine estuary? 167
8.7. Conclusion 169
8.8. References 170

**Chapter 9. Metropolitan Trajectories for Anticipatory
Governance of Urban Biodiversity** 175
Charles DE GODOY LESKI and Yohan SAHRAOUI

    9.1. The challenges of an attractive city faced with ecological
    injunctions: contextual elements of emerging governance............ 176
    9.2. The cognitive stakes of a collaborative territorial prospective 180
    9.3. Scenarios of metropolitan trajectories: contrasted
    political–ecological footprints 183
        9.3.1. Strategic scenario: "Dense City" or the Return of the
        Rhine Model (Scenario 1)................................. 186
        9.3.2. Dystopian scenarios: "city–nature opposition" (Scenario 2)
        and "city–nature interweaving" (Scenario 3) 189
        9.3.3. Utopian scenarios: "radical ecological restoration"
        (Scenario 4.1) and "optimal reconciliation" (Scenario 4.2) 189
        9.3.4. Transformative scenario: resilient city (Scenario 5) 191
    9.4. Conclusion 192
    9.5. References 193

Conclusion... 197
Denis SALLES, Glenn MAINGUY and Charles DE GODOY LESKI

List of Authors 203

Index 205

About the Authors

Gilles Billen is Emeritus Director of Research at CNRS, attached to the UMR Metis at the University of Sorbonne, France. His research focuses mainly on agro-food systems from a biogeochemical perspective.

Henrique Cabral is Director of Research at INRAE, Cestas. He was a full professor at the Faculty of Sciences of the University of Lisbon (FCUL), and Director of the Marine and Environmental Sciences Centre (MARE) and the Oceanography Centre in Portugal. His current research focuses on the impacts of global change on estuarine and coastal systems.

Cécile Capderrey is a coastal engineer at BRGM in Orléans, France. She is a member of the Coastal Risk and Climate Change research unit.

Jeanne Dachary-Bernard is a research fellow in economics at INRAE, center for Bordeaux, in the ETTIS research unit. She works on the influence of environmental risks, on individual residential choices and on adaptation to climate change.

Alain Dupuy is Professor of Hydrogeology at the Bordeaux Polytechnic Institute (Bordeaux INP), France and has been Director of ENSEGID-Bordeaux INP (*École nationale supérieure en environnement, géoressources et ingénierie du développement durable*) since 2013. He is a member of the EPOC Laboratory within the PROMESS team, where he coordinates research on pressure, mass and energy transfers in multilayer aquifers.

Michael Elliott is Professor of Estuarine and Coastal Sciences at the University of Hull, UK, and was the Director of the former Institute of Estuarine and Coastal Studies (IECS). His teaching, research and consultancy activities focus on estuarine and marine ecology, policy, governance and management.

Josette Garnier is a research director at CNRS and coordinates the UMR Metis biogeochemistry team at the University of Sorbonne, France. From 2007 to 2018, she directed the Ile-de-France Federation of Environmental Research (FIRE), which brings together 18 laboratories from a wide range of disciplines. She is a member of the French Academy of Agriculture. Her research activities cover a wide range of topics, from the biogeochemistry of agro-ecosystems to the ecological functioning of aquatic environments in the land–sea continuum.

Charles de Godoy Leski is a CNRS contract researcher at the Laboratoire de géographie physique (UMR 8591) and a research associate at the Centre Émile Durkheim (UMR 5116), France. His research focuses on the socio-cognitive processes of public environmental action in the planning and construction of territories.

Julia Le Noë completed her PhD thesis at the University of Sorbonne, France on "Biogeochemical Functioning and Trajectories of French Territorial Agricultural Systems. Carbon, Nitrogen and Phosphorus Fluxes (1852-2014)". She is currently attached to the Institute of Social Ecology at the University of Natural Resources and Life Sciences in Vienna, Austria, where her work focuses on forest transitions and land use change.

Mario Lepage is a research engineer in aquatic ecology at INRAE Nouvelle-Aquitaine Bordeaux, France. He is a member of the Aquatic Ecosystems and Global Change research unit, and part of the team charged with studying the functioning of estuarine ecosystems.

Glenn Mainguy holds a PhD in sociology. He is a research associate at the Centre Émile Durkheim (UMR 5116), France. His research focuses on the analysis of major contemporary transformations – from socio-economic mutations to socio-ecological transitions in France and Russia – and on the unequal ways in which social groups experience these upheavals.

Thierry Oblet is a lecturer at the Faculty of Sociology at the University of Bordeaux and a researcher at the Centre Émile Durkheim (UMR 5116), France. His research work has focused on social and urban policies for the past 30 years.

Yohan Sahraoui is a lecturer in Geography at the University of Burgundy Franche-Comté and a researcher at the ThéMA Laboratory (UMR 6049), France. His research work focuses on socio-environmental changes related to territorial planning, using participatory modeling and popular education approaches.

Denis Salles is Director of Research at ETTIS-INRAE in Bordeaux, France. He directs research in the sociology of the environment and public policies on the anticipation of climate change on territories and on water governance.

Nabil Touili is a contractual teacher-researcher in land use planning at the University of Versailles Saint-Quentin-en-Yvelines, Université Paris-Saclay, France. His research and teaching focuses on risk governance and interdisciplinary studies.

Jean-Paul Vanderlinden is a professor of ecological economics and environmental studies at the University of Versailles Saint-Quentin-en-Yvelines, the Université Paris-Saclay, France and the University of Bergen, Norway. His research and teaching focus on emerging socio-technical risks, adaptation to climate change and interdisciplinary practices.

Florian Vergneau is an economist and data science consultant for Veltys. A graduate of the Toulouse School of Economics (TSE), France, his work focuses on competitive bidding models and econometric tools as applied to financial transactions.

Aude Vincent is a contract researcher in hydrogeology at the University of Iceland (Reykjavik) and an alumnus of the École normale supérieure, France. Her work concerns the impact of human activities on groundwater, directly (irrigation) or indirectly (climate change), especially in coastal areas (risk of seawater intrusion) and near glaciers.

Acknowledgments

This book is the result of a collective research collaboration carried out between 2017 and 2020 as part of the interdisciplinary URBEST project: *Shaping adaptive governance in estuarine cities: Bordeaux Métropole and Gironde estuary facing global change.*

URBEST has brought together expertise in sociology, economics, political science, ecology and hydrogeology to study the challenges of adaptation to climate change and global change faced by cities located on estuaries, known as "estuarine cities". Due to the specificity of their issues, estuarine cities are challenged to imagine adaptive and anticipatory governance for their socio-ecological transition.

URBEST was financially supported by LabEx COTE and the IdEx of Bordeaux[1]. The symposium on "Adaptive governance and global change in estuarine cities" (*Gouvernance adaptative et changements globaux dans les métropoles estuariennes*), organized on October 17, 2019 at the University of Bordeaux, aimed to present and compare the main results of URBEST with researchers who have studied other configurations of estuarine metropolises, and with socio-economic and political actors and civil society of the Bordeaux Métropole, the city concerned by the research. The book brings together all the contributions of the URBEST project. We would particularly like to thank the researchers who took part in the URBEST

---

1 This study was funded by the ANR, the French National Research Agency, within the framework of the PIA, the future investing program, hosted by *Laboratoire d'excellence COTE* (ANR-10-LABX-45) and IdEx Bordeaux.

project: Pieter Leroy, member of the LabEx COTE scientific council, and Gilles Pinson from the Centre Émile Durkheim. URBEST was hosted and supported from 2015 to 2019 by the ETTIS research unit of the INRAE – Centre Nouvelle-Aquitaine Bordeaux (Environment and land use, transitions, infrastructures, societies).

Introduction

The world's largest cities (Tokyo, New York, Buenos Aires, New Orleans, Bangkok) and 10 major cities in Western Europe are located near coastlines and estuaries. The expansion of coastal cities is emerging as a major trend in the 21st century. It is expected that by 2060, 1.7 billion people will be located along coastal areas. These cities are now faced with the challenges of climate change and, more broadly, global change[1]. Historically, coastal and estuarine cities were highly attractive both demographically and economically, due to the opportunities offered by their strategic location for mobility, economic exchanges and quality of life. At the same time, they are subject to strong social, economic and ecological vulnerabilities, in terms of risks, such as: erosion of the coastline, marine submersion and flooding of dense urban areas; in terms of urban heat islands, water-stressed environments (severe low water levels) or the concentration and diffusion of urban contaminants.

Introduction written by Denis SALLES, Glenn MAINGUY and Charles DE GODOY LESKI.

1 The term "global change" refers to biophysical and societal processes that, from the global to the local, affect the living conditions and the habitability on Earth. They are made up of the combined effects of anthropogenic environmental crises linked to climate change (warming, extreme events, etc.), the erosion of biodiversity (pollution, degradation of biotopes, depletion of resources, etc.), the globalization of socio-economic exchanges, demographic dynamics and migratory mobility, food requirements, and the sustainability of access to natural resources (deforestation, depletion of fossil fuels, problems of freshwater access, pressure on fishery resources) (Blondel 2008; Salles 2013).

For several estuarine cities of the North Atlantic (London, Bilbao, Bordeaux, etc.), the renewal of their city centers since the 2000s has involved the outsourcing – more often imposed than chosen – of traditional port activities outside the city boundaries. The requalification of former industrial areas, the development of a tertiary economy, urban planning based on real estate construction for commercial, tertiary and residential purposes, and the display of an environmental dimension and a quality living environment have formed the political basis for the transformation of these cities. These processes of metropolization have resulted in an intra-metropolitan concentration regarding the mobilization of natural, economic, demographic, technical and land resources.

The dynamics of metropolization has been accompanied by a concomitant structural process of metapolization, which refers to "the set of spaces in which all or some of the inhabitants, economic activities or territories are integrated into the daily (ordinary) functioning of a city" (Ascher 1995, pp. 35–36, *author's translation*). The peripheral territories of estuarine cities, already marked by the decline of traditional rural and estuarine activities (port industries, agriculture, shellfish farming, fishing, hunting, etc.), have been confronted with issues of pressure from peri-urban or tourist urbanization, exposure to flood risks, or spaces designated as flood expansion zones to protect the dense city (Sautour and Baron 2021). Estuarine communities must also face the supervision of their future development through urban planning regulations (Flooding prevention plan, European Water Directive Framework). Due to their ecological interest, these estuarine areas are concerned by policy tools for biodiversity conservation via ecological corridors (the "green-blue" framework of Grenelle de l'environnement). Their potential for the development of tourist or recreational activities (such as the growth of river cruises) is highly sought after by territorial projects.

Finally, from an institutional point of view, the French estuarine territories (metropolitan areas and their estuarine hinterlands) were subject to new inter-municipal responsibilities in terms of risk management and the environment, in particular the management of aquatic environments and protection against flooding.

These socio-spatial configurations, which combine metropolization/ metapolization, exposure to global changes and climate risks, as well as territorial reforms, are coupled with social and political tensions that revise new forms of territorial conflict/cooperation (Laurent 2021; Pinson 2021). This social resistance has been embodied in recent years by an increase in territorial conflicts (Subra 2018) against the implementation of facilities (Mele 2013), or by a rise in explicit criticism within professional sectors (especially agricultural) or social movements – such as that of the yellow vest movement since 2018, protesting in particular against social and environmental territorial inequalities (access to resources and public services, mobility infrastructures, environmental norms) (Mermet and Salles 2015; Deldrève et al. 2021).

If the image of "peripheral France" (Guilluy 2014) reflects on the interests and logics specific to urban or rural territories in separate and antagonistic ways, global changes participate in a new, more integrative territorial reading of the issues, abolishing the boundaries between different typologies of spaces (Löw 2015). Understanding global changes in estuarine cities requires shifting the urban question to a broader territory: the city and its hinterlands. Looking beyond the boundaries of the city puts the interplay between the urban and its environment back at the center of the analysis. It is becoming necessary to take into account the continuities that make up the territory and the political–ecological imprint of urban spaces in the peripheral territories. The interdependent relationships between Bordeaux Métropole and the Gironde estuary, much like those of the cities of the Loire and Seine estuaries, are emblematic of the way socio-ecological configurations are torn between urban attractiveness and ecological vulnerabilities in the face of climatic challenges and global changes.

The thesis developed in this book is that the urgency of global change challenges estuarine cities to imagine anticipatory governance. This governance must take into account the socio-ecological interdependencies woven into the fabric of the city and the protection of the environment. It must also support a territorial solidarity extended to all estuarine areas.

Based on this explicit consideration of socio-ecological interdependencies, what adaptation trajectories are being outlined for estuarine cities in the face of global change? What anticipatory scenarios are emerging to build the urbanities of tomorrow? Through the collective production of worldviews, the integrative, contributive and critical research proposed in this book aims to contribute to this socio-ecological transition.

This book, composed of nine contributions from different disciplines (territorial biogeochemistry, ecology, economics, hydrogeology, geography, political science, sociology), proposes an integrated reflection centered on the case study of the estuarine city of Bordeaux and in addition on the results observed on the estuaries of the Loire and the Seine.

The first part of this book interconnects hydrogeology with sociology to interpret the stakes of global changes from the controversy study aimed at substituting drinking water resources by rerouting freshwater from rural catchment areas to the city.

In Chapter 1, "The Governance of Socio-Ecological Interdependencies: The *Landes du Médoc* Water Catchment Area Controversy", Charles de Godoy Leski explores the socio-ecological interdependencies between cities and their estuarine hinterlands in terms of drinking water supply. This chapter proposes a socio-historical analysis of the socio-technical, ecological and political controversy that has surrounded the Landes du Médoc water catchment project since its emergence about 20 years ago. It successively explores the historic construction of the ecological vulnerability of the drinking water supply of Bordeaux Métropole, the uncertainties of the scientific advances in the hydrogeological modeling of water resources and requirements, and the controversies on the potential ecological impacts of such withdrawals on forestry.

In Chapter 2, "Ecological Engineering in a Controversial Drinking Water Production Project", Alain Dupuy and Aude Vincent ask the following question: When groundwater extraction is inevitable to supply the city, what types of actions could avoid both societal concerns and possible environmental impacts? The authors explore technical avenues to circumvent a classic for-and-against situation, exploring ways to mitigate the potential impacts of the Landes du Médoc groundwater project in Gironde. They present a solution inspired by artificial aquifer recharge (AAR) techniques that requires no additional infrastructure.

The second part of this book addresses the challenges of global change for estuarine cities through the analysis of flood risk management. Three particular interpretations are developed through:

– land transactions;

– ecological engineering solutions;

– multiple social meanings attached to these risks.

In Chapter 3, "Living in a City Exposed to Flood Risk: At What Cost(s)?", Jeanne Dachary-Bernard and Florian Vergneau examine the impact of flood risk on the residential logic at work in estuarine territories. What are the socio-economic mechanisms that govern the spatial organization of land and, among them, does vulnerability to flood risk play a structuring role? The authors apply the hedonic pricing method over the period 2011–2016 to the territory of the city, extended to its neighboring estuarine municipalities. They show that the risk management mechanisms studied have a double effect on real estate prices.

In Chapter 4, "The Ecological Restoration of Estuaries: Protection of People and Combating the Erosion of Biodiversity", Mario Lepage, Michael Elliott, Cécile Capderrey and Henrique Cabral analyze various experiences of ecological restoration in estuaries (the Gironde estuary in France, the Mondego estuary in Portugal, the Scheldt estuary in Belgium). Based on the observation that estuarine environments have been confronted for several decades with a growing tension between the attractiveness of such territories and a specific vulnerability linked to their position as an interface between land and sea, and in view of the many functions they offer, they show how restoration of the ecological quality of these environments has become a real challenge in terms of protecting people and combating the erosion of biodiversity.

In Chapter 5, "Sensemaking in the Face of Estuarine Flood Risk Mitigation", Jean-Paul Vanderlinden and Nabil Touili explore the types of interpretations and sensemaking attached to climate change, and in particular to flood risk. The authors explore the nature of the frames of reference that make sense to actors faced with flood risk mitigation options. They demonstrate that the meaning-making process iteratively invokes frames of reference associated with causality, salience and moral norms.

The third part of this book explains the historical relationship between cities and estuaries, and their consequences.

In Chapter 6, "The Estuarine City as an Allegory for Changes in Solidarity", Thierry Oblet presents the estuarine city as an allegory for observable contemporary transformations in the ways of thinking on the notions of interdependence and solidarity. His account is based on the history of Bordeaux and its estuary, as a privileged field of observation for challenges associated with the transition from a solidarity – in the 20th century, within the framework of state governance – based on the recognition of the interdependence of individuals within a nation, to a solidarity linked – in the 21st century, within the framework of globalization and metropolitan governance – to the recognition of the interdependence of territories. The author explains that the estuarine city is more than just a model for organizing the cooperation between the territories of Gironde; it reveals all its complexity and rough edges.

In Chapter 7, "Nantes and the Loire: Construction of an Estuarine City Faced with Port and Environment Challenges", Glenn Mainguy analyzes the way in which institutional actors participate in making visible the socio-ecological interdependencies between an estuary (that of the Loire) and an urban center (Nantes) in the context of global change. The author studies how the estuarine dimension of the city and the territory of Nantes is embodied in discourses and practices.

The fourth part of this book is devoted to the modes and methods of anticipation developed to imagine the future of estuarine cities in the context of global change.

In Chapter 8, "Past and Future Socio-Ecological Pathways of the Seine Estuary", Gilles Billen, Julia Le Noë, Camille Noûs and Josette Garnier retrace, from the middle of the 19th century, the major stages in the evolution of the estuary and its trajectory of artificialization, based on the analysis of material flows linked to the commercial traffic of the ports of Rouen and Le Havre. They show how the history of the development of the estuary is entirely conditioned by that of its watershed. In the case of the Seine, the trajectory is induced by increasing centralization and urban polarization: Paris must be connected to the rest of the world in order to maintain its position as a *World City*. The use of a long-term perspective allows the authors to describe two future scenarios (up to 2050) of the socio-ecological system of the Seine.

In Chapter 9, "Metropolitan Trajectories for Anticipatory Governance of Urban Biodiversity", Charles de Godoy Leski and Yohan Sahraoui explore the ways in which biodiversity is linked to urban policies in order to think about the future of Bordeaux Métropole. They mobilize the results of a collaborative foresight exercise, including various stakeholders of the city (researchers, local authorities, government services, NGOs, etc.), which aimed to develop scenarios on biodiversity in the city by 2035. The authors describe five divergent metropolitan trajectories to anticipate a future that could potentially integrate urban development with concerns on biodiversity preservation.

References

Ascher, F. (1995). *Métapolis ou l'avenir des villes*. Odile Jacob, Paris.

Blondel, J. (2008). Les changements globaux. *Forêt Méditerranéenne*, 19(2), 119–126.

Deldrève, V., Candau, J., Noûs, C. (eds) (2021). *Effort environnemental et équité. Les politiques publiques de l'eau et de la biodiversité en France*. Peter Lang, Bern.

Guilluy, C. (2014). *La France périphérique*. Flammarion, Paris.

Laurent, E. (2021). La métropole coopérative. Éléments d'analyse et de mesure. *Les conférences POPSU* [Online]. Available at: http://www.urbanisme-puca.gouv.fr/IMG/pdf/discourslaurent_web_v2_1_.pdf.

Mele, P. (ed.) (2013). *Conflits de proximité et dynamiques urbaines*. PUR, Rennes.

Mermet, L. and Salles, D. (eds) (2015). *Environnement : la concertation apprivoisée, contestée, dépassée ?* De Boeck Supérieur, Louvain-la-Neuve.

Pinson, G. (2021). Les métropoles au chevet de la décentralisation ? *Les conférences POPSU* [Online]. Available at: http://www.urbanisme-puca.gouv.fr/les-metropoles-au-chevet-de-la-decentralisation-a2537.html.

Salles, D. (ed.) (2013). Enjeux sociétaux, vulnérabilités face au changement climatique. In *Les impacts du changement climatique en Aquitaine. Un état des lieux scientifique*, Le Treut, H. (ed.). Presses universitaires de Bordeaux, Bordeaux.

Sautour, B. and Baron, J. (eds) (2021). *L'estuaire de la Gironde : un écosystème altéré ? Entre dynamique naturelle et pressions anthropiques.* Presses universitaires de Bordeaux, Bordeaux.

Subra, P. (2018). *Géopolitique de l'aménagement du territoire*. Armand Colin, Paris.

PART 1

The Water of the Cities and the Water of the Fields

1

The Governance of Socio-Ecological Interdependencies: The Landes du Médoc Water Catchment Area Controversy

1.1. Introduction

The supply of drinking water to metropolitan areas has long been considered a local issue inserted in a network based on territorial interdependencies. After guaranteeing universal supply lines in the 1980s, water utilities are now faced with the challenge of securing the quantity and quality of water resources and their long-term sustainability (Barbier and Roussary 2016).

The combined effect of the demographic and economic attractiveness of cities, and the exposure of rural areas to climate change (Salles and Le Treut 2017) frames drinking water supply in a new light. Bordeaux Métropole, which obtains its drinking water from deep aquifers that are subject to strong anthropic and climatic pressures, has been engaged for the past 20 years in a decentralization process and inter-territorial interactions to diversify its supply sources.

The city needs water to anticipate its demographic growth. Access to new resources, located outside its institutional perimeter, requires negotiations, agreements and compensation, which are sources of territorial tensions within a metapolitan space whose democratic authorities have difficulty governing the extensive spatial, social and economic interdependencies (Ascher 1995).

Chapter written by Charles DE GODOY LESKI.

The issue of water supply for human consumption (drinking water) highlights the interdependence between Bordeaux Métropole and its functional rural areas of the Landes du Médoc (LdM) catchment area. This issue has gradually become controversial since, on the one hand, Bordeaux Métropole has considered access to these underground resources as a substitute for the overexploitation of the Eocene water table for its drinking water supply and, on the other hand, local communities – particularly forestry representatives – have contested a detour or even a monopolization of water for urban purposes at the expense of the forestry and ecological functions of the local territory.

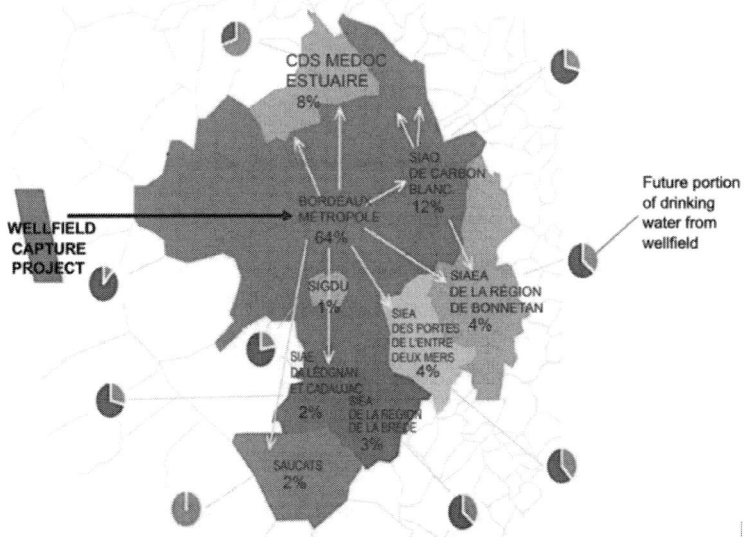

CDS MEDOC ESTUAIRE	Intermunicipal Medoc Estuary Organization
BORDEAUX MÉTROPOLE	Bordeaux Métropole Public Authority
SIAO DE CARBON BLANC	Carbon Blanc Water Supply Organization
SIGDU	Emergency Management and Information System
SIAEA DE LA RÉGION DE BONNETAN	Intermunicipal Organization for Drinking Water Supply and Sanitation Management of Bonnetan
SIEA DES PORTES DE L'ENTRE DEUX MERS	Intermunicipal Organization for Drinking Water and Sanitation Management of Portes Entre Deux Mers
SIEA DE LEOGNAN ET CADAUJAC	Intermunicipal Organization for Drinking Water Supply in Léognan and Cadaujac
SIEA DE LA RÉGION DE LA BRÈDE	Intermunicipal Organization for Drinking Water Supply in La Brède Region

Figure 1.1. *Map showing territorial interdependencies through the substitution mechanism of the Landes du Médoc water catchment area to the Bordeaux metropolitan area. For a color version of this figure, see www.iste.co.uk/salles/estuarinecities.zip*

This chapter proposes a socio-historical analysis of the socio-technical controversy surrounding this infrastructure project since its emergence some 20 years ago. It successively explores:

– the historical vulnerability of the water supply of the Bordeaux Métropole;

– the uncertainties of scientific advances in the hydrogeological modeling of water resources and requirements;

– local controversies on the potential ecological impacts of water withdrawals on silviculture.

In fine, this study reveals the new configuration that socio-ecological interdependencies are creating in the relations between metropolises and their hinterlands.

How are environmental effects anticipated by scientific knowledge and then grasped by the actors of water resource governance? How does the visibility of socio-ecological interdependencies between Bordeaux Métropole and the Landes du Médoc reconfigure governance for the drinking water issue? What are the political narratives that accompany this infrastructure project and its critiques?

In Gironde, the history of the socio-technical regimes (Geels and Schot 2007) of water is intertwined with the history of the city and the history of interdependence with the resource territories near and far from the urban centers that consume water. The socio-technical regime is a relational system between institutions and social and political norms, which generate rules and procedures that regulate its stability. The dynamics of change are most often exercised through incremental innovations, linked to territorial components (Guyot Phung and Charue-Duboc 2020), such as local scientific specialization in the study and exploitation of deep aquifers. The modalities of transfer or circulation of knowledge between actors are decisive for initiating transitions (Geels 2002). These are required not only by existing metropolization dynamics but also, in a more fundamental way, by the ecological and climatic interdependencies that link territories in metropolitan areas (de Godoy Leski 2021).

1.2. Drinking water supply in Gironde, the history of a transfer from surface water to deep groundwater

While the development of hydrogeological techniques has allowed access to groundwater, the recurring issue of water scarcity remains an issue that is actualized throughout local political and technical history (Lorrain and Poupeau 2014). The Gironde still has diverse and abundant water resources (rivers, estuaries, streams, lakes, man-made water bodies, groundwater and deep aquifers). The history of the exploitation of the volumes that can be extracted and the contamination of surface resources has punctuated the more or less distant (re)locations of metropolitan catchment points in order to supply the city with drinking water.

1.2.1. *Under the Roman Empire: the administration of a city's water sources in its estuary*

Bordeaux owes its location in part to the springs of drinking water initially located near the gravel terrace of Mont Judaïque, which prompted the first settlers to choose this site as it was favorable to trade due to its navigability by river and estuary. Burdigala is a city "open" to other territories near and far including Tolosa, following the conquest of the Narbonne at the end of the second century BCE. Enlarged territorial interdependencies were gradually established through trade routes, which followed the maritime and river routes. In around the 2nd century, for a population estimated at 20,000 inhabitants spread over 125 hectares, water needs were met by local springs and wells. The precise location of the sacred spring Divona[1] is not known, upon which the construction of Burdigala, then an emerging city and the result of Gallo-Roman urbanism, was based. Divona, a goddess venerated by the Gauls, gave her name to many springs regarded as miraculous, and to which the 4th-century Latin poet Ausonius[2] proclaims the eternal splendors:

---

[1] Otherwise known as the Bouquière spring or the Saint Christoly spring. The history of urbanism by the patrimonialization of these names, in streets or in places, provides some clues as to their perimeter, and is the basis for assumptions on the localization of these mythical water points through architectural and urbanistic memory.

[2] Born in 309/310, either in Bazas or Bordeaux, and died around 394 in his villa located between Langon and La Réole, he is considered at the same time as a man of letters, a politician and a Gallo-Roman pedagogue of the Later Roman Empire.

Salve, fons ignote ortu, sacer, alme, perennis,
Vitree, glauce, profonde, sonore, inlimis, opace!
Salve, urbis genius, medico potabilis haustu,
Divona, Celtarum lingua, fons addite divis!³

Among the hypotheses put forward by historians of Bordeaux (Auguste Bordes, Camille Jullian), the Devèze would take its name from it. Other sources are cited: Audège, La Daurade, Mirail. Gallo-Roman urbanism bears the mark of the sophistication of the socio-technical regime of the period, through the construction of hydraulic works that transported water to public fountains, thermal baths and private homes. The awareness of the sanitary risk and water contamination was solved through a system of sewers responsible for evacuating contaminated water away from residential areas. The Gallo-Roman gravity-fed distribution system collected water from a higher elevation in the city to guarantee the supply circuit. This technical constraint led the Romans to seek water from further and further away in order to meet the demographic growth of an appealing city. The historical socio-ecological interdependencies between Burdigala and the resource territories outside the original boundaries of the city underline the structuring character of the socio-technical regime of water for the urban ecology of Gironde.

1.2.2. The Middle Ages: the era of hydraulic and defensive withdrawal of the city

Despite its location far from the mouth of the Gironde estuary, what the inhabitants of Burdigala feared most was an onslaught by the Germanic tribes on Roman Gaul, which led to fortifications being built. The widespread looting and destruction of cities accelerated the urban transition from an open city into the fortified city of Castrum. By the year 900, the demographic growth could no longer be contained by the fortifications and urban expansion forced the city to spread past its ramparts. The port infrastructures were subject to this change, particularly in terms of new builds to secure the city. An inner harbor was built on the Devèze, secured and controlled by a water gate. The urban planning of water supply lines was weakened through conflicts and the destroyed aqueducts were never rebuilt.

---

3 "Hail, fountain of source unknown, holy, gracious, unfailing, crystal-clear, azure, deep, murmurous, shady, and unsullied! Hail, guardian deity of our city, of whom we may drink health-giving draughts, named by the Celts Divona,—a fountain added to the roll divine!" (translation by Hugh G. Evelyn-White).

Until 1520, the inhabitants obtained their water supply from local springs and through an urban network of wells, tapping into the water table for consumption. At the end of the 15th century, the city of Bordeaux was already an important city for which drinking water supply was once again a problem directly linked to its demography, the original quality of the resource and the impact of human activities that were highly concentrated within the city walls.

1.2.3. *The hydraulic "Renaissance" in the 17th century: towards the golden age of Bordeaux fountains*

When adopting the long-term view to explain the socio-ecological dynamics that structured the city of Bordeaux and its territorial expansion, the history of Bordeaux's fountains raises questions about the relationship of the inhabitants to water, to nature and to urbanization. The transfer of hydraulic techniques between cities is reflected in the first water conveyance completed in 1520 by Jean Comilhot, master fountain builder for the city of Rouen. The project aimed to link Bordeaux with a network of fountains: Brachet, Maurian and Mirail. A network whose utilitarian functions of drinking water supply are combined with playful and aesthetic urban amenities in the public space. The interdependencies between the city and the services rendered by nature favored the emergence of a department of fountain workers, initially created for the maintenance of the Bouquière spring. The doctrine established for this office was mainly to ask nothing from the inhabitants who used the fountains to wash their clothes. The right to water is closely related to the rights to the city. Modernizing the social through the urban (Oblet 2005), by relying on the socio-technical regime of a given era, translates an essential social function of cities: to offer access to the technical innovations of an era in order to incorporate them into the lifestyle of its inhabitants. At that time, water, considered a requirement of the urban community, was free at the fountains; only the delivery to houses by water merchants was priced[4]. However, all of these sources were of poor quality and insufficient in quantity to satisfy the daily needs of the population. With the shift of secondary uses from fountains to streams and ditches, health problems emerged as a consequence of the discharge from

---

4 Concerning fire hazards, the supply of water for the fire department was a non-financial obligation imposed on the water merchants in exchange for the Figueyreau and Lagrange springs given as a concession to this corporation – a local legal status that would end in 1858.

contaminants emitted by commercial activities of an urban system (Dupuy 1984) connected to the surface water system. The water from the La Fontaine de l'Or, initially intended to supply the La Grave district, became cloudy in the rainy season and quickly stagnated. The unfortunate sailors who carried this water in their barrels grieved for the curative and mystical virtues declaimed by the poet Ausonius and suffered harmful sanitary effects. However, the public problem was not so much drinkability as the quantity of water available to meet the increasing needs of urban growth. Indeed, the socio-technical regime of this period was not able to build prospective knowledge on the state of the resource in the face of an accelerated anthropization of the world.

1.2.4. *From 1800 to 1850: the dark age of resource availability*

The city of Bordeaux did not change its water supply system much during the first half of the 19th century, despite the growth of its population. During this period, the city experienced a water shortage. As early as 1820, measures were taken by the urban government of Bordeaux. From then on, water for street cleaning was rationed every summer. While the urban dynamics of river and estuary cities are architecturally translated through the implementation of a network of bridges to efficiently connect urban territories dispersed through the surface hydrographic network, the political–ecological imprint of these structures on the demand for water remains to be explored. The construction of the Pont de Pierre in 1822 put pressure on the right bank as a new space for urbanization. In 1829, the municipality of Bordeaux gave up watering the green spaces of the Jardin Public and the Allées de Tourny. Faced with this critical socio-ecological situation, competitions and calls for tenders were launched. The actors of the socio-technical regime of the time were split into three camps. The confrontation raged between the supporters of spring water exploitation, those of the Garonne waters and those who promoted artesian wells. The available water was polluted, unhealthy and the risk of epidemics was permanent for the 130,000 inhabitants of Bordeaux in 1850. The insalubrity of surface water elevated the exploitation of deep spring techniques at the forefront of engineering. This historical memory of local shortages still informs the territorial rationalization of water actors:

In 1850, in Bordeaux, the drinking water service only allowed 5 liters per person per day... This is the equivalent of a third world country today (Director, SMEGREG, *translated by author*)[5].

However, water needs are historically defined by the institutionalization of norms and ordinary practices. Indeed, the use of water in 1950 differs from contemporary uses, promoting hygiene standards and access to running water.

1.2.5. *From 1850 to 1950: towards extractivist geopolitics of water*

At the end of the 1870s, the shortage and poor quality of water led the public authorities to look for new resources in distant territories; especially given that the wells within the city were running dry and the availability of drinking water was estimated at 3.5 liters per inhabitant per day. The search for quality water ruled out local solutions: drilling artesian wells, diverting the Jalle de Blanquefort or filtering water from the Garonne. In 1878, on the proposal of the engineer Wolf, working for the city of Bordeaux, the City Council turned its attention to the Fontbanne springs in Budos. The socio-technical regime of this period placed great importance on geographical slopes in order to transport water by gravity. The political decision was based on leveling studies between Budos and Bordeaux, which confirmed the technical exploitation of this resource, some 40 kilometers away from the southeast of the city. An aqueduct, built between 1885 and 1887, carried the water from the Fontbanne spring to the Béquet station in Villenave-d'Ornon[6]. The Budos aqueduct, which is still in operation, provides Bordeaux with a capacity of 28,800 m^3/day. The local historian Jean Dartigolles never ceased to point out that "Budos was wronged" (*translated by author*). In 1888, the mayor took the matter to court and summarized the situation during a municipal council meeting: "The canal was built, the water diverted, the city of Bordeaux satisfied and the community of Budos deprived of its water" (*translated by author*). This

---

5 Syndicat mixte d'étude et de gestion de la ressource en eau du département de la Gironde, a syndicate for the study and management of the Gironde water resources.
6 With a length of 42 km and crossing 15 communes, the Budos aqueduct owes its implementation to the topographical characteristics which offer a difference in level of 0.1 mm/m, that is, a gentle slope of 4 m from the starting point to the finishing point.

memory of local history feeds a still current dispute on the pricing of the cubic meter of drinking water in Budos. With a cost of 7.50 euros, it is more than double the price of a cubic meter in Bordeaux Métropole, which is 3.60 euros. The contemporary controversy of the Landes du Médoc water catchment area feeds on the analogy of the choices and social effects of the socio-technical regime of an era attitude to territorial resources.

1.2.6. *From 1950 to the present day: the socio-technical anticipation of issues through the governance of the drinking water resource*

The public drinking water supply crisis of the Bordeaux agglomeration begins with the question addressed in 1956 to the Prefect of Gironde by Professor Schoeller[7] (1899–1988) on the recharge time of the deep aquifers already being exploited for drinking water in Bordeaux. This eminent Bordeaux hydrogeologist alerted not only his emerging scientific community but also the public authorities to the risks of saline intrusion in the deep aquifers of the Gironde, due to potentially excessive exploitation of groundwater (Schoeller 1956)[8]. A decade later, on January 1, 1968, the Bordeaux Urban Community (CUB in French) was formed. Management of the drinking water supply became one of its competences, even if it did not cover the whole of its political–administrative territory. Between 1968 and 2013, the growing population of the CUB (+194,665 inhabitants in 2021 according to INSEE), through the spread of suburban housing (70% of single-family homes built in the Gironde), required the connection of homes to the water network, with a peak between 1970 and 1975, and an end to connections in 1984.

The problem of drinking water re-emerged on the political agenda during the 1990s under the political impetus of the Gironde departmental council. The scientific answer to the questions of hydrogeologist Schoeller was

---

[7] Teacher-researcher at the Geology Laboratory of the Faculty of Sciences of Bordeaux, author of numerous fundamental treatises in hydrogeology, then an emerging discipline, noteworthy are *Hydrogéologie* (1955); and *Les eaux souterraines. Hydrogéologie dynamique et chimique. Recherche, exploitation et évaluation des ressources* (1962).

[8] Method to determine the extraction radius of boreholes and the Darcy coefficient as applied to the study on the depletion of the Aquitaine water table through the Paleocene sands: Schoeller (1956).

finally acknowledged only in 1995[9]. This scientific answer to the long temporality is partly explained by the formalization of the theories and scientific tools of the then young discipline of hydrogeology, which developed in the shadow of geology:

> The answer came in 1996, 40 years later: yes, there is a risk of saline intrusion. That is to say, between the time he asked the question and the time the answer came, there was all the science that produced the theorization of how deep aquifers work (Director, SMEGREG, *translated by author*).

The political response was formalized 63 years later with the signing, on January 24, 2019, of the financing contract for the "Landes du Médoc" wellfield project by the main institutional actors involved in the governance of drinking water resources in Gironde.

In Gironde, the exploitation of underground resources is exposed to the risk of salinization of the Eocene water table in coastal and estuarine areas. This risks of saline intrusion in drinking water catchments in a superficial environment has steered the history of science and technology the Bordeaux university site, leading to the development of institutions such as BRGM[10] Nouvelle-Aquitaine or the ENSEGID[11] school of engineers, organizations specialized in the study of subterranean resources.

The transfer of technical progress from the exploitation of geological petroleum resources to the exploitation of groundwater opened up access to the resources of deep aquifers. Thanks to these advances, the historical problem of a lack of drinking water in the Gironde region was considered to have been solved.

---

9 Synthesis of knowledge on the Eocene water table in Gironde, Dordogne and Lot-et-Garonne prior to the establishment of a SAGE: BRGM (1995). *Synthèse des connaissances sur la nappe de l'Éocène en Gironde, Dordogne et Lot-et-Garonne en préalable à l'établissement d'un SAGE*. In the present case, the analysis is limited to the Gironde department, and subject to the requirements of the SAGE Nappes profondes.

10 Office of geological and mineral resources: *Bureau des ressources géologiques et minières*.

11 National graduate school specializing in environment, georesources and sustainable development engineering: *École nationale supérieure en environnement géoressources et ingénierie du développement durable*.

However, the ecological and economic implications of the exploitation of the deep aquifers of Gironde continue to form part of the contestation of the expertise led by different scientific partners of the project of the Landes du Médoc water catchment area (BRGM, INRAE[12], ENSEGID):

> A particularity of water management, if you talk to me about surface water management, drought or flooding... Anybody who goes on the bridge, sees that the water is no longer flowing, will be able to see for themselves that there is a drought; if the water is overflowing, they will see that there is a flood; if the fish are belly up, they will see for themself that there is pollution. When it comes to groundwater, the average person does not have direct access to the report, it always goes through the filter of the expert (Director, SMEGREG, *translated by author*).

The production of agreements on the meaning of action (present and future) cannot be carried out without making explicit the knowledge constructed and mobilized by the different institutions. The normative and cognitive framing of this institutional work is necessary for the appropriation, contestation and evolution of knowledge in decision-making bodies.

1.3. Metropolitan territorial conceptions in the face of forestry references

The political reference to the Landes du Médoc territorial entity is an old term contested by the foresters, by virtue of the peri-urbanization of this territory contributing to the retreat of forest areas. The many names for this resource relocation project reveals its "troubled" temporality: "*Oligocène de Sainte-Hélène*", "*champ captant de Sainte-Hélène*", "*grand projet opérationnel*", "*grand projet de ressource de substitution*", "*projet de champ captant des Landes du Médoc*". The expression used in official documents and websites for this major project remains the "Landes du Médoc water catchment area" (*champ captant des Landes du Médoc*), 12 of the 14 potential drilling points of which are located in the Saumos community. In Gironde, this spatial dynamic is accentuated along the railway axis linking

---

12 National research institute for agriculture, food and the environment: *Institut national de recherche pour l'agriculture, l'alimentation et l'environnement*.

the cities of Bordeaux-Arcachon-Libourne (BAL), which initially saw the "Landes de Bordeaux" disappear in favor of home ownership by the working classes in the 1960s and 1970s; this was followed in the 1990s by the middle and upper classes in the western part of the agglomeration, in Pessac and Mérignac and then in a second phase, by the extension of suburban housing outside the borders of the metropolis in the Landes du Médoc.

The peri-metropolitan populations express doubts and criticisms about the capacity of the institutions of Bordeaux Métropole to inform and protect them from climatic and ecological risks. Taking into account the effects of global change on the peri-urban forests of the city of Bordeaux is inseparable from the acceleration of metropolitan needs in terms of land, residential spaces, environmental resources (water, air, biodiversity) and the living environment that concern the ecosystems near and far from the centers of activity. The forests of the Landes de Bordeaux, according to their historical name, have been managed and exploited since Napoleon III in a highly and historically anthropized environment.

On December 22, 2014, an information meeting organized by the inter-CLE of the SAGE Nappes profondes[13] was held at the Sainte-Hélène community village hall, which has since become a place for expression for opponents of the LdM wellfield project. Provision 88 of SAGE states that the inter-CLE "ensures regular exchanges with the CLEs of other territories" (*translated by author*) involved in these major projects. This ad hoc process brings together the presidents of the CLEs of the SAGE Nappes profondes and the SAGE Lacs Médocains, concerned by the potential impacts of the LdM project on the drying up of the Médoc lakes.

The joint mobilization of economic, climatic and hydrogeological knowledge aims to quantify and avoid the risk of "drying out" the roots of maritime pines in the Landes du Médoc:

> Simulation tools are the only way to share the findings. In hydrogeology, the only way we have to share things is to do modeling (Director, SMEGREG, *translated by author*).

---

13 The interdependence between the boundaries SAGE produces on systematic exchanges between the facilitators and the CLEs. The inter-CLE is a formal deliberative body that has existed since 2014 and has met seven times to deal with the Landes du Médoc wellfield project. See: https://participation.bordeaux-metropole.fr/participation/developpement-durable/champ-captant-des-landes-du-medoc.

Inter-CLE can also be understood as an informational and communicational tool (Lascoumes and Le Galès 2004), since this approach:

> [...] aims to ensure that local stakeholders are well informed of the ins and outs of the project, to identify the questions raised by its implementation and to ensure that answers to these questions are provided as soon as possible[14].

The cognitive work around this project starts from the observation of an interdependence between the city and its functional territories as a mechanism through which to influence metropolitan action. The opening of new governance apparatuses (inter-CLE) makes visible the socio-ecological interdependencies with which the territorial actors must engage in for any deals linked to the project.

During the 2014 inter-CLE presentations of the project, controversy arose first with the foresters, according to whom, surface drainage techniques are deemed essential to the preservation of the maritime pine monoculture. This historical surface drainage technique is now potentially counterproductive to long-term groundwater recharge conditions, which requires that stormwater be allowed to infiltrate the soil and flow through the various geological layers.

The frequency and intensity of rainfall cause high-intensity stormwater to flow into streams. Low- and medium-intensity rainfall allows water to infiltrate. This controversy over the risk of dewatering the pines has galvanized the social tensions surrounding the Landes du Médoc wellfield project.

From a hydrogeological point of view, the maximum volume that can be extracted from a deep aquifer is a function of the extraction rate, the hydrodynamic characteristics of the aquifer tapped, as well as of the tapping method. A significant amount of digital modeling work was necessary in order to simulate scenarios representative of past, present and future extraction methods. The construction and calibration of the hydrogeological model and the construction of potential abstraction scenarios are two complex tasks that both require a specific application and are subject to uncertainties specific to the modeling work, hence the importance to further

---

14 Excerpt from the first convocation letter dated November 3, 2014, signed by both presidents.

refine the mesh of the models in order to stabilize the results according to potential drilling locations.

Confined aquifer

Piezometric surface

Reservoir ceiling

Reservoir

Drawdown without dewatering

Dewatering of the aquifer

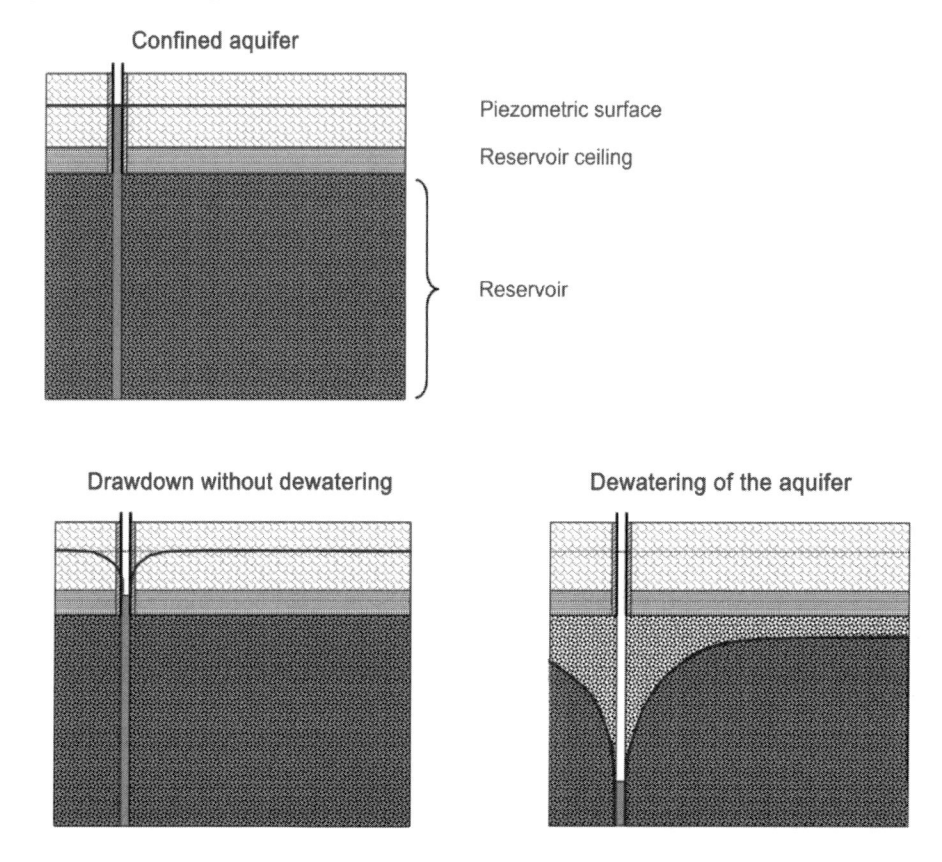

Figure 1.2. *Technical dimension of the dewatering of an aquifer in the surface environment (source: SAGE Nappes profondes (2012)). For a color version of this figure, see www.iste.co.uk/salles/estuarinecities.zip*

Based on these models, the presentations of the BRGM hydrogeological studies have given rise to a hostile reception from local actors within the CLE of the SAGE Lacs Médocains, and the local residents. The solutions for the supply of drinking water to Bordeaux Métropole and the allocation of catchments for the water supply of urban areas are seen as an interference with their territory. When the pre-consultation announced by the project owner was launched, several oppositions emerged. In January 2015, the silviculturist union exposed "the uncertainties about the project's impacts on the production of maritime pines and their fear of a strong lowering of the so-called 'sand' water table, which is essential to forestry" (inter-CLE, *translated by author*). The mayors of the communities concerned have also expressed their opposition:

> As far as I'm concerned, and even if this is imposed on us, I will issue a very unfavorable opinion of this project, and I know that the mayors of the neighboring communes are in the same frame of mind (Mayor of Saumos 2015, *author's translation*).

Forest owners express the feeling of being caught between the urban area of influence and the ongoing development of the coastline:

> In the part of the Médoc where we are, we are somehow stuck between the galloping development of a city that wants to be a millionaire in number of inhabitants as soon as possible – if we have understood correctly – and on the other hand the development of the coastline that also brings constraints, while we represent more traditional activities – that doesn't mean that they are not modern – and we have the feeling of being a little bit neglected and stuck (Silviculturist, SYSSO, *author's translation*).

The inhabitants expressed doubts about their communal political representation in the territorial geopolitics involved in this project:

> The Mayor is an employee of Bordeaux Métropole, so one might wonder if there is not a conflict of interest. Then there was a public meeting on the PLU: while all the other communities are developing in terms of population, here, we are tightening the screw, we are not developing, forbidden to develop (Resident of Saumos, *author's translation*).

This territorial mistrust is historically accentuated by the withdrawal of the State from the territories, in particular the end of public engineering as of 2007, and by the difficulty of the rural world to draw collective benefits from metropolization. This last point reflects the difficulty of enacting strategies for territorializing public action into a social system, which is not denied by Act III of the decentralization act, which focuses on cities to the detriment of their hinterlands. Not to mention the socio-political formalization of a general interest that has yet to be refined – territorial differentiation, otherness and solidarity:

> They [the institutional holders of the project] called on the population to come and learn about this subject, their concerns and their outcries because, understandably, they have not been reassured, so inevitably they are against it. It is logical: their questions are not answered and, once again, it is the human condition to be afraid for one's land, for one's culture. It is logical. As long as there is doubt about the effects of this wellfield, there will be a lot of opposition (Elected municipal official, Le Temple, *author's translation*).

This mistrust underlines the importance of learning mechanisms in the ongoing development of territorial competencies and in the acculturation to new decision-making methods in the face of uncertain knowledge regarding climate change. The dividing lines can be observed between elected officials, between elected officials and technicians and between technicians and stakeholders in water governance[15].

In fine, it is the territorial learning mechanism structured around the inter-CLE that has the function of circulating normative and cognitive elements that are supposed to promote support for the project. The negative reception of "quantitative indicators" can be explained in part by the domination of expert knowledge (such as the models criticized for their uncertainties) over the vernacular knowledge shared by the residents of the Landes du Médoc:

> Everything is justified by numbers, so we have a million inhabitants versus 500 inhabitants; economic interests versus what they imagine as an empty forest. I believe that we have ecosystem services (water, etc.) that could argue for a less hierarchical relationship between Bordeaux Métropole and a small village like Saumos. We need more solidarity, something

---

15 These concern the *Syndicat des sylviculteurs du Sud-Ouest* (SYSSO) – the largest union of forest owners in France – the *Société pour l'étude, la protection et l'aménagement de la nature dans le Sud-Ouest* (SEPANSO), the inhabitants of the towns affected by the project's impacts, forest owners and silviculturists in all their diversity (the notion of sector is to be put into perspective here).

that is at the same level of hierarchy in the relationships (Saumos resident, *author's translation*).

The persistence of scientific obstacles in the robustness of models in the face of climatic uncertainties places the hydrogeological expert in an uncomfortable situation regarding their role as mediator between science and a society expecting sanctified knowledge:

> The expert is going to have very specific and partial data about the resource, and yet they will be obliged to interpret it, to say, 'The resource is in such and such a state' (Director, SMEGREG, *translated by author*).

However, while uncertainty is the hallmark of hydrogeological modeling, it is accused of arbitrariness by project opponents as soon as this expertise is mobilized to legitimize a political decision. Hence the importance, in this process of knowledge circulation, of a driving actor (Raulet Croset 2017) who frames funding, compensation guarantees and technical and political experimentation. Several institutions have held this function, but in turn, there was a collective expectation that Bordeaux Métropole would be the official project manager by 2019.

1.4. The refinement of models and the rising criticism

One of the main stumbling blocks in the controversy over the Landes du Médoc wellfield concerns the validity of the hydrogeological and hydro-climatic models that have been employed for several years in an attempt to objectify the state of the Gironde groundwater resource.

Since 1997, specific modeling has been carried out on the pilot site of the future LdM wellfield by integrating new meshes in the hydro-dynamic model MONA[16] until 2012, then later in the Phonème model, which has undergone three successive versions, in February and then October of 2015 and again in 2018. The three versions of the Phonème model have been parameterized on an extraction estimated between 10 and 12 million cubic meters over a time span of 10 to 12 years.

---

16 In the tradition of regional models developed in the 1970s, the North Aquitaine Model (MONA in French) developed by BRGM was initially designed in response to the collapse of the Eocene water table in the Gironde. This evolving model supports public policies, hence its exposure to social protest.

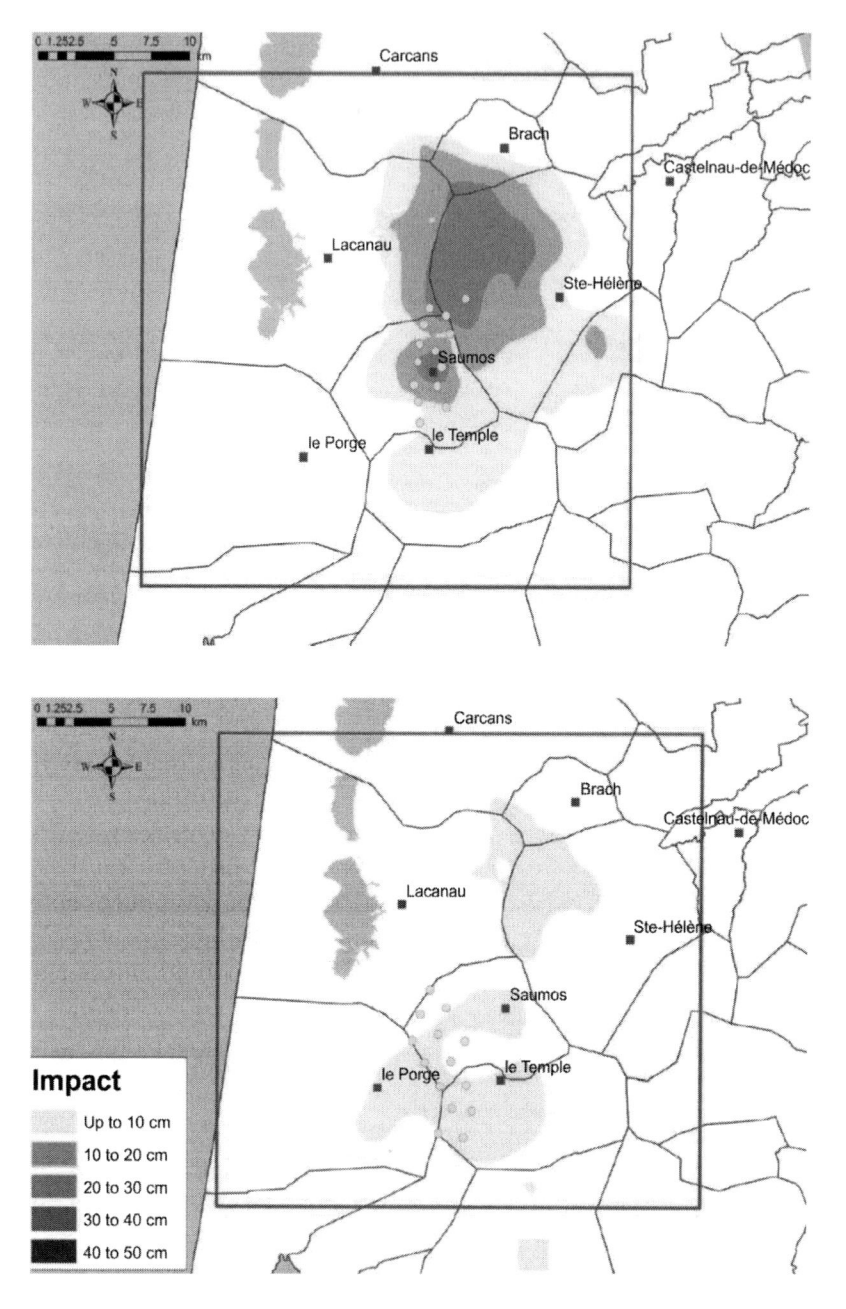

Figure 1.3. *Evolution of the maximum impact on the environment (Phonème V3): overestimation after 10 years of operation (source: BRGM (2018)). For a color version of this figure, see www.iste.co.uk/salles/estuarinecities.zip*

COMMENT ON FIGURE 1.3.– *Note for the reader: as part of its modeling work, BRGM studied the influence of ecological and climatic impacts according to the pressure exerted on the environment by the distribution of the 14 boreholes in the project. The map above presents the results of the model with a maximum impact of 30 cm on the surface environments, and the last configuration retained presents impacts limited to a drop in the water level necessary to maintain forest ecosystems evaluated at 10 cm.*

The simulations of the Phonème[17] Version 1 model were presented by BRGM in the inter-CLE consultation arenas on February 27, 2015. The impact of the extractions was associated with a *potential* risk of lowering surface water levels *estimated* at 1.70 m on the surface environments. The results of these first digital simulations were used as references for the validation of the project and to determine the borehole locations.

Version 2 of the Phonème model, presented on October 1, 2015, increased the impact estimate to 1.40 m. Version 3 of Phonème in 2018, built to refine the sensitivity of the simulations, estimated the impact at 40 cm. In order to limit the impact in time and space, several borehole configurations were simulated by the Phonème model.

Figure 1.3 shows the evolution of the results between the position initially studied (top map) and the reference state presented during the public utility inquiry in 2021–2022 (bottom map). The scientific response aims to limit the effects of the maximum drawdown in the Plio-Quaternary water table, which feeds the pine trees and the biodiversity of the Landes du Médoc territory. Between the initial version of the wellfield and the reference state, the impacted surface area has been halved and the maximum impact has been divided by 3. We can observe that the location of the 14 borehole points – initially concentrated in the commune of Saumos – has been moved westwards and distributed between the communes of Le Porge, Le Temple and Saumos.

In order to further reduce uncertainties, in 2018, the results of Phonème V3 were coupled with the GO+ hydroclimatic model developed by INRAE. The GO+ model incorporates climate variables to refine the estimate of the

---

17 Unlike MONA, the Phonème model was created specifically to measure the impact of the installation of the LdM wellfield on silviculture following concerns expressed during the official unveiling of the project in 2014.

wellfield's impact on primary maritime pine production over a 20-year operation. This process modeling is used to describe and simulate the biophysical and biogeochemical functioning of forests. In view of launching the public utility inquiry in 2021–2022, these results aim to optimize the location of the wellfield and validate the position of the boreholes and their safety perimeter.

Fears concerning the impacts of water extractions on the superficial environment, and the forest in particular, are expressed by the joint voices of silviculturists, inhabitants and the association *Vive la forêt*[18], whose president is a retired economist and researcher at CNRS. The social pressure of fears, relayed by local associations and unions, has also focused on the urban demographic projections of Bordeaux Métropole and the rural projections of coastal towns. Centered on the development of Bordeaux, they are accused of underestimating the water needs related to residential and tourist activities in the coastal zone (Lacanau). These demographic calculation hypotheses, noted during the presentation of the initial models at the inter-CLE, have been modified.

The modeling studies initially conducted by BRGM, ENSEGID and INRAE had the political and social vocation of providing scientific arguments that were sufficiently robust in order to make the large-scale project for a substitute resource in the Landes du Médoc legitimate and acceptable. On the other hand, it is clear that despite the continuous and cumulative effort of expertise, the controversy has intensified, fueled by a questioning of the scientific validity of the expertise. Scientists have found themselves at a loss when faced with a surfeit of criticism that counter the assertion made by successive simulations of a reduction in the impact on the water table, and therefore the threat to forestry activities.

"We are used to communicating our scientific results to individuals who trust us" (*translated by author*), the fatalist remarks of a scientist observing the supplanting of the debate on values of trust rather than on arguments of scientific validity.

With the explicit integration of the climate change dimension, evapotranspiration was considered as an input variable of the Phonème V3 model by estimating its effects on forest production. The simplified equation

---

18 http://www.vivelaforet.org/.

calculates the effects of global warming based on the assumption of a +3% to +5% increase (depending on whether the sky is overcast or not) in potential evapotranspiration (PET) per additional degree of temperature (+1° = +3% to +5% PET).

After the integration of climate variables into the model, there was a shift in criticism towards the form of the experts' presentations: "We don't understand anything. The slides are not clear, there would be no slides and it would all be the same" (Silviculturist, *author's translation*). The limits of model coupling were highlighted during each inter-CLE by the difficulties of "controlling the uncertainties resulting from the cumulative effects of variables" (modeling scientists, *author's translation*).

In the context of this social construction of uncertainty, while the experts claim a better objectification of the impacts, the local populations suspect an underestimation or, worse, a dissimulation of the risk. The governance of such a project would require a reduction in the overarching and prescriptive role of the experts, and the involvement of local studies and surveys by silviculturists, which have been discredited up until now. The relations of social and technical reciprocity tend to reconfigure the system of actors, and thus the socio-technical regime of water, through the expression of different tensions (Monstadt and Coutard 2019) within the inter-CLE. Associating and communicating on the difficulties of forecasting, even making it a discursive dimension integrated into the major project of alternative resources, would promote, if not a convergence of visions, at least the restoration of a climate of trust. All the more so since the spatial and ecological dimension of this project is a reminder, if one were needed, that rural communes and their groupings are organized differently from urban territories (Béhar et al. 2015).

1.5. Conclusion

The construction of what is at stake regarding the sustainability of the deep groundwater of the central Eocene, intended for the supply of drinking water to Bordeaux Métropole and the Gironde department, is confronted with tensions between various economic, scientific, institutional and political conceptions. The socio-ecological interdependencies brought to light by this issue feed new ecological and political conceptions, and proceed from a rethinking of territorial asymmetries in the interactions between an urban consumption territory and a rural resource territory. The recognition of interdependencies is played out politically within the framework of deliberative water bodies (inter-CLE), but this visibility of those interdependencies has only emerged following a controversy, or even a territorial crisis (Toubin et al. 2013).

The construction of a sustainability issue for the deep aquifers of the central Eocene is thus placed in tension between economic, scientific, institutional and political interdependencies. The main issue is now the collective search for trajectories likely to respond to the territorial challenges posed by global changes. Concepts on the sustainability of deep aquifers politically re-engage territorial solidarities between communities and between urban and rural ecosystems.

In this deliberative framework, this public controversy illustrates cognitive tensions between actors who do not share the same references and the same visions of the territory. It is a tension between, on the one hand, vernacular knowledge claimed to be robust by the experience of the silviculturists, but which is made invisible by scientific expertise, and, on the other hand, expert knowledge that has scientific legitimacy, but which is socially and politically contested by the local residents. Social structures (lifestyles) and territorial structures (attractiveness of urban centers) have been invited into the arenas of debate concerning the evolution of urban services (Maresca 2017). However, rural populations raise doubts and criticisms about the capacity of the public institutions in charge of the project to inform and protect them from climatic and ecological risks. The silviculturists contest the input hypotheses of the Phonème model and the evocation of the results, far from corresponding to the image of a science that is progressing in its methodologies, but which arouses distrust and mistrust by the uncertain nature of a scientific result presented as a proof of evidence by public action. The territorial conception of the links between social and hydrogeological dynamics is based on expert appraisals whose study sites are defined by the economic interdependencies of water: cost-sharing mechanisms between exploitation, interconnection and rerouting to attractive urban territories.

The normative and cognitive framing of trust is necessary, not only for the circulation of knowledge, but above all to the explanation of issues intertwining social and ecological solidarities in the (re)composition of worlds (Descola 2014). Therefore, in December 2019, during the annual conference organized by Bordeaux Métropole precisely on the theme of cooperation, Henri Sabarot, president of the CLE of the SAGE Lacs Médocains and president of the recent PNR Médoc, concluded his intervention by underlining the tensions linked to the wellfield project. He reminded the audience that it was "a problem of form rather than substance", and that "technocracy was not the right level at which to discuss these issues" (*author's translation*). This is a way of going beyond the limits of a socio-technical regime that is too often a refuge from politics. The politicization of interdependencies produces new strategies and new scientific and deliberative experiments. However, the production of agreements on the meaning of interdependencies (present and future) cannot be exercised without making explicit the knowledge constructed and mobilized by the various institutions.

1.6. References

Ascher, F. (1995). *Métapolis ou l'avenir des villes*. Odile Jacob, Paris.

Barbier, R. and Roussary, A. (2016). *Les territoires de l'eau potable. Chronique d'une transformation silencieuse (1970–2015)*. Quae, Versailles.

Béhar, D., Levy, J., Beja, A., Padis, M.O. (2015). Y a-t-il une bonne échelle locale ? *Esprit*, 96–108.

Bonneuil, C. and Fressoz, J.-B. (2016). *L'événement Anthropocène. La Terre, l'histoire et nous*. Le Seuil, Paris.

BRGM (2018). Modèle Phonème : apport des investigations de terrain. Évaluation de l'impact du champ captant des Landes du Médoc. Report, BRGM/RP-68406-FR, Orléans.

Descola, P. (2014). *La composition des mondes. Entretiens avec Pierre Charbonnier*. Flammarion, Paris.

Dupuy, G. (1984). Villes, systèmes et réseaux. Le rôle historique des techniques urbaines. *Les annales de la recherche urbaine*, 23/24, 58–71.

Fernandez, A. (2003). L'économie municipale à Bordeaux, XIXe–XXe siècles : les mutations de l'édilité. *Histoire, économie et société*, 22, 413–436.

Geels, F.W. (2002). Technological transitions as evolutionary reconfiguration processes: A multi-level perspective and case study. *Research Policy*, 31, 1257–1274.

Geels, F.W. and Schot, J. (2007). Typology of sociotechnical transition pathways. *Research Policy*, 36(3), 399–417.

de Godoy Leski, C. (2021). Vers une gouvernance anticipative des changements globaux. L'emprise des interdépendances socioécologiques sur une métropole estuarienne. Bordeaux Métropole et l'estuaire de la Gironde. PhD Thesis, Université de Bordeaux, Bordeaux.

Guyot Phung, C. and Charue-Duboc, F. (2020). La dimension territoriale : modalités d'émergence et de diffusion de la niche sociotechnique. *Finance Contrôle Stratégie* [Online]. Available at: https://doi.org/10.4000/fcs.4942.

Lascoumes, P. and Le Galès, P. (2004). *Gouverner par les instruments*. Presses de Sciences Po, Paris.

Lorrain, D. and Poupeau, F. (2014). Ce que font les protagonistes de l'eau : une approche combinatoire d'un système sociotechnique. *Actes de la recherche en sciences sociales*, 3(203), 4–15.

Maresca, B. (2017). Modes de vie : de quoi parle-t-on ? Peut-on le transformer ? *La pensée écologique*, 1(1), 233–251.

Monstadt, J. and Coutard, O. (2019). Cities in an era of interfacing infrastructures: Politics and spatialities of the urban nexus. *Urban Studies*, 56(11), 2191–2206.

Oblet, T. (2005). *Gouverner la ville. Les voies urbaines de la démocratie moderne*. Éditions PUF, Paris.

Raulet Croset, N. (2017). De la coopération à l'organisation territoriale émergente : à la jonction des situations, des espaces, et des activités. HDR Thesis, Université Panthéon-Sorbonne, Paris.

SAGE Nappes profondes de Gironde (2012). Plan d'aménagement et de gestion durable de la ressource. Tome I : Synthèse de l'état de lieux et de l'analyse économique. Exposé des enjeux. Synthèse des orientations de gestion, SMEGREG, Commission locale de l'eau, Conseil départemental de la Gironde.

Salles, D. and Le Treut, H. (2017). Comment la région nouvelle-aquitaine anticipe le changement climatique ? *Sciences, eaux & territoires*, 1(22), 14–17.

Schoeller, H. (1956). Méthode de détermination du rayon d'appel des forages et du coefficient Darcy, application à l'étude de l'épuisement de la nappe des Sables paléocènes de l'Aquitaine. *Ass. Intern. d'hydrogéologie scientifique*, 41, 67–75.

Toubin, M., Diab, Y., Laganier, R., Serre, D. (2013). Les conditions de la résilience des services urbains parisiens par l'apprentissage collectif autour des interdépendances. *Vertigo*, 13(3).

2

Ecological Engineering in a Controversial Drinking Water Production Project

Although the strong interactions between human activities and the hydrological cycle have long been recognized (Massuel et al. 2018), publications addressing them have increased since 2012 (Blair and Buytaert 2016) with the emergence of socio-hydrology as a research field in its own right (Sivapalan et al. 2011). Socio-hydrology views humans and their activities as endogenic to the water cycle, rather than external factors and/or boundary conditions (Sivapalan and Blosch 2015). Its goal is to understand the dynamics and co-evolution of coupled "human–water" systems (Sivapalan et al. 2011). The socio-hydrological approach uses models with hydrological variables and "human behavior" variables (Blair and Buytaert 2016): either including a social component in a hydrological model or creating an interface for actors to test the effects of their choices in hydrological simulations. However, human behavior modeling remains very limited and simplified, remaining quite far from the effectiveness of hydro(geo)logical models (Westerberg et al. 2017).

In an attempt to overcome these limitations, and in order to increase the actionability of interdisciplinarity to meet many expectations (Wesselink et al. 2016; Barthel et al. 2017), some authors have proposed additional tools for an effective socio-hydrological approach. On the one hand, they combine scholarly knowledge with local knowledge. For example, they combine a borehole owner's historical and present experience with their hydrological knowledge (Riaux and Massuel 2014). In other cases, local knowledge of the

---

Chapter written by Alain DUPUY and Aude VINCENT.

connections between surface and groundwater hydrology enhances the understanding of hydrogeologists (Massuel et al. 2018). On the other hand, they develop a field approach to interdisciplinarity. They share common ground with researchers from different disciplines in order to agree on spatial and temporal scales, to produce a common vocabulary and representations, and finally to choose angles of analysis that make sense for all subjects involved (Riaux and Massuel 2014). Finally, they practice reflexivity at each stage of the research (Jollivet and Legay 2005).

In the context of climate change and water stress, one technique now being considered is artificial groundwater recharge (AGR) of aquifers for management purposes. AGR is also referred to as "groundwater replenishment", "water banking", "sustainable groundwater storage", "enhanced recharge" or "artificial recharge" (Dillon 2005). This intentional aquifer recharge technique is intended for future water recovery or environmental benefit (IAH[1]; Dillon 2005). The water used may be from a natural source or from appropriately treated wastewater. An improvement in groundwater quality may also be an expected gain (IAH; Dillon 2005).

AGR has many advantages, including low cost and good durability (IAH; Dillon 2005). The use of this technique is now expanding rapidly. For example, AGR is widely promoted in southern India, primarily through the rehabilitation of *tanks*, small reservoirs typically created by damming an intermittent stream (Vincent 2007; Massuel et al. 2014). It is also implemented through infiltration basins in Spain (Grau-Martínez et al. 2018) and injection wells in Namibia (Murray et al. 2018). The conditions necessary for successful AGR action depend on existing hydraulic infrastructure and space for water harvesting (Dillon 2005). Of course, AGR technology is not a cure for overexploited aquifers (Dillon 2005). Without proper study and management of the technique, the downstream basin may suffer from water depletion (Dillon 2005).

2.1. The socio-hydrogeological configuration of the Landes du Médoc catchment area (Gironde)

The study area is located in southwestern France in the Gironde department, close to the metropolis of Bordeaux. The climate of the area is

---

1 https://recharge.iah.org/.

temperate and oceanic (Loustau et al. 2016). Precipitation is frequent throughout the year, and more significant in autumn and winter. A water deficit can occur from April to September. Evapotranspiration is significant in summer and is a controlling factor in free aquifer level variations (Roux 2006; Naldeo 2015). It is estimated that in the New Aquitaine region, climate change is expected to accentuate current trends and create stress on the resource at low-flow periods (AcclimaTerra and Le Treut 2020).

The domestic water supply of Bordeaux is mainly extracted from the deep Eocene aquifer, which is already slightly overexploited, and which has justified the search for new alternative water sources. The exploitation of the Oligocene aquifer is being considered in the St. Helena area to replace part of the existing domestic water supply for the Bordeaux area and local villages. The projected extraction of 10 mm^3/year would replace two thirds of the current extraction for drinking water from the Eocene aquifer. A well field, that is, a group of wells and boreholes drawing water from the same aquifer, should be established in the central Médoc in the near future. Several kilometers from the ocean, the area is mainly covered with maritime pines, cultivated for their wood, and initially planted under Napoleon III to drain the marshes. There are also a few vineyards, as well as some industry.

The construction of the wellfield has raised concerns and controversy among the local population and environmental associations (see Chapter 1). Even though the simulated impact on the non-captive aquifer, located well above the captured formation, is low (less than 30–40 cm maximum lowering of the water table), could the pine forest and biodiversity suffer from water deprivation/scarcity? The local biodiversity is specific to the area that used to be a swamp. This area has been drained by numerous canals and ditches since the 19th century. These maritime pine plantations constitute the Landes de Gascogne Forest, the first cultivated forest massif in Europe (Saltel et al. 2016). This area, with its ecological originality, is located in the Médoc Regional Nature Park created in 2019.

The extraction of water from a deep confined aquifer can cause a lowering of the water level in the upper unconfined aquifer. As a result, surface ecosystems and vegetation are at risk of suffering from water deficits, and even more so going forward as climate change conditions evolve. According to the ERC (*"Éviter, Réduire, Compenser"*: Avoid, Reduce, Compensate) doctrine, one approach would be to avoid extraction, another to minimize it and another to compensate for the effects of the withdrawal.

This chapter focuses on the ecological compensation option and its potential consequences. It presents the results of numerical tests of the method for mitigating the impacts of groundwater extraction in deep aquifers on surface systems.

The potential impacts of extraction in the confined aquifer are seen in a lowering of its own confined water table (a), but potentially also in a lowering of the water table in the unconfined aquifer (b) (Figure 2.1).

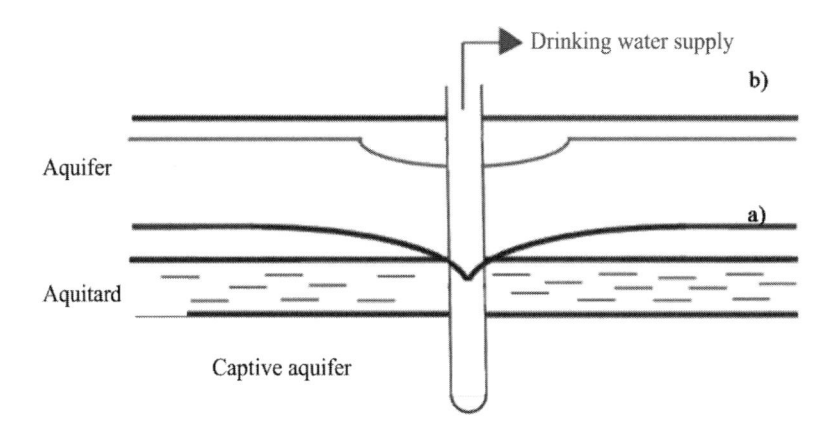

Figure 2.1. *Potential impacts of water collection in the confined aquifer. For a color version of this figure, see www.iste.co.uk/salles/estuarinecities.zip*

To compensate for this potential impact, an environmentally friendly engineering solution inspired by artificial aquifer recharge (AAR) techniques and requiring no additional infrastructure was considered. Specifically, this solution is based on the principles of aquifer storage and recovery (ASR) and aquifer storage transfer and recovery (ASTR), which involve injecting water into a well for storage to achieve subsequent water recovery (Dillon 2005).

2.2. An ecological engineering solution

Therefore, the proposed method is to reinject a small amount of the water extracted from the confined deep aquifer into the surface systems through existing drainage ditches or shallow wells. Certain conditions must be considered for any AGR project: space for water collection, which does not apply in our case since the water used for reinjection is pumped from the confined aquifer and existing water infrastructure[2]. The proposed engineering solution was tested with a simple heuristic model developed with the Modflow® ModelMuse interface. The actual local geology was simplified by considering, from bottom to top, the Oligocene aquifer, an aquitard overlying several geological layers, and the unconfined aquifer in the Plio-Quaternary formations (based on descriptions, for example, in Vigneaux (1975), BRGM (1997), EGID (2001) and Roux (2006); see Table 2.1). The different parameters applied were thicknesses (average of actual thicknesses) and hydraulic coefficients (hydraulic conductivities and storage coefficient) based on previously conducted pumping tests (BRGM 1997; EGID 2001, 2003; Roux 2006) and other models developed in the region (MONA: Pédron et al. 2009; Phonème: Saltel and Arnaud 2015) (see Table 2.1).

	Thickness (m)	K (m/d)	S_s (-)	S_y (-)
Plio-Quaternary	45	43.2	0.1	0.38
Aquitard	12	0.004	1.10^{-5}	0.2
Oligocene	57	0.5	1.10^{-8}	0.25

Table 2.1. *Table of the key parameters used in the heuristic model*

The recharge estimated using the Modflow® Recharge (RCH) Package, by a classical approach for modeling recharge (Boble and Crosbie 2017), is an annual average of 36.5 mm/year (BRGM 1997).

Subsequent studies (BRGM for Phonème) set the other boundary conditions by a model built and used to estimate, with the highest possible accuracy, the potential impact of the wellfield on hydrogeological dynamics (Saltel and Arnaud 2015). The alternative boundary conditions are thus set to:

– no flow for the northern and southern boundaries (i.e. nothing is applied), as in the Phonème model (Saltel and Arnaud 2015);

---

2 These are numerous within the study area and easy to reuse, which is a strong point.

– a specified head of 0 m (Time-Variant Specified-Head (CHD) Package) for the western boundary, representing the influence of the ocean, as in the Phonème model (Saltel and Arnaud 2015);

– a specified head of 10 m (Time-Variant Specified-Head (CHD) Package) for the eastern boundary, representing the existing piezometric crest delineating the groundwater catchment in both aquifers (EGID 2003; Pédron et al. 2009).

Existing ditches, draining the open water table, are represented using the Modflow® Drain Return (DRT) Package, which allows water to exit the system only when the water table reaches the bottom of the ditch. All tests first include a preliminary steady-state phase, followed by transient phases set according to the tested reinjection time. In order to simulate the well field, an extraction of 10 mm^3/year is applied through the well flow distributed over 32 km^2. In order to simulate water reinjection from the wells, the Modflow® Well Package is used with positive values. In order to simulate water reinjection via ditches, the Recharge (RCH) Package on a square with sides a few meters long is used to inject the desired precise volume, combined with the Stream (STR) Package to then distribute the reinjected water. After a run, the stream leakage-in/stream leakage-out balance check needs to be performed to neutralize the Stream (STR) Package: distribution without addition or withdrawal.

Since the goal is to compensate for wellfield-induced drawdown on the unconfined aquifer, the results are presented as a percentage of sustainable drawdown, calculated relative to the unmitigated wellfield impact. A result of 100% sustainable drawdown would mean that the mitigation method has no effect. The goal needs to be as close as possible to a 0% sustainable drawdown.

2.3. How much of the extracted water must be reinjected?

The results were calculated for different drawdown thresholds: 5, 10, 15 and 20 cm. The impacts on surface systems for drawdowns below 10 cm are assumed to be negligible. Three sets of sensitivity tests were performed on:

– the percentage of water reinjected;

– the duration of the reinjection;

– the chosen method to reinject the water.

The optimal solution estimates the amount of extracted water to be reinjected at 5%, with reinjection starting on the same day as extraction for both reinjection methods. A total of 25 trials were conducted.

2.3.1. *Percentage efficiency of the water reinjected*

For each percentage of water reinjected, simulations were run (Figure 2.2) for different reinjection times and with all the different reinjection methods and configurations. Obviously, if no water is reinjected, the sustainable drawdown is 100%.

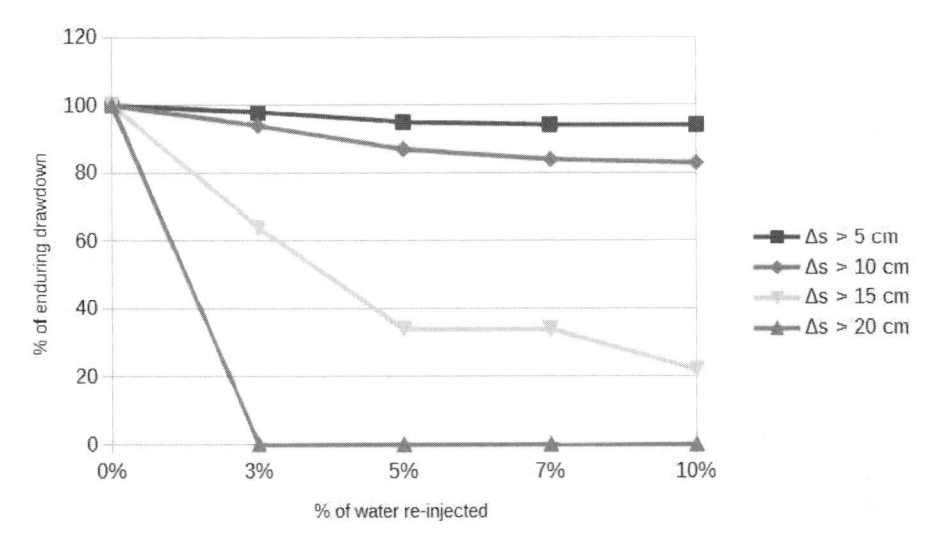

Figure 2.2. *Percentage of permanent drawdown in the unconfined aquifer as a function of the percentage of extracted water reinjected, shown for drawdowns of 5, 10, 15 and 20 cm. Results shown are the average number of runs with different methods for a full year of reinjection (four runs for each percentage of reinjection tested and a reference run (no reinjection), 17 runs). For a color version of this figure, see www.iste.co.uk/salles/estuarinecities.zip*

With 3% of the extracted water reinjected, the attenuation effects are of moderate magnitude. With 5% of the extracted water reinjected, the results are significantly better, since two thirds of the drawdowns greater than 15 cm are avoided. With 7% or even 10% of the extracted water reinjected, the additional benefit is small. Therefore, 5% of water reinjected seems to be optimal for our purposes.

2.3.2. *Efficiency of the reinjection duration*

It turns out that the difference is significant and that an early start for reinjection maximizes attenuation efficiency. Reinjection delayed by 4 or 6 months significantly reduces the quality of the results.

2.3.3. *Efficiency of the selected method to reinject the water*

The third and final sets of tests dealt with the method and configuration chosen to reinject water – through drains or wells – as well as their number. No significant differences were found (Figure 2.4), and no clear optimum emerged from the results.

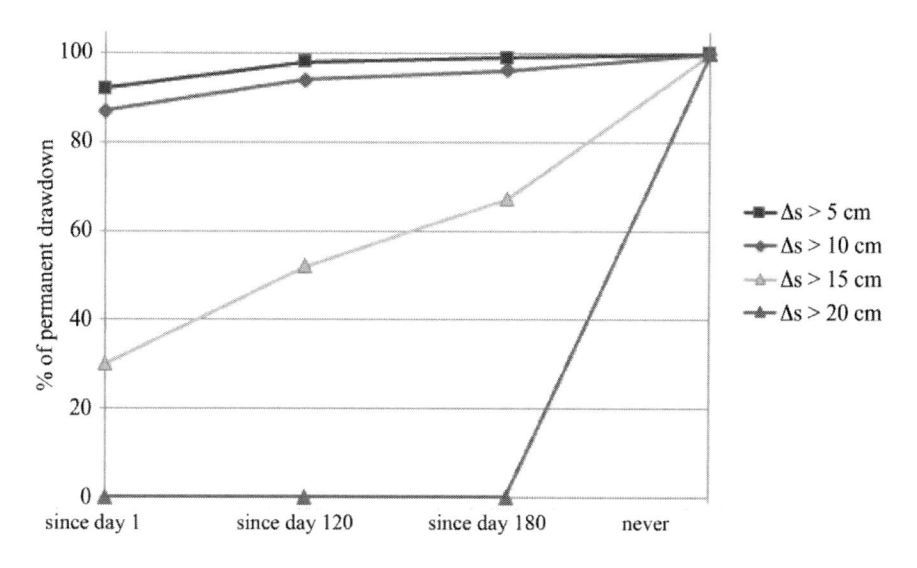

Figure 2.3. *Percentage of permanent drawdown in the unconfined aquifer as a function of percentage of reinjection time, shown for drawdowns of 5, 10, 15 and 20 cm. Results shown are the average number of tests for different reinjection methods and 5% reinjected water (four tests for each reinjection period tested and one reference test (no reinjection), 17 tests). For a color version of this figure, see www.iste.co.uk/salles/estuarinecities.zip*

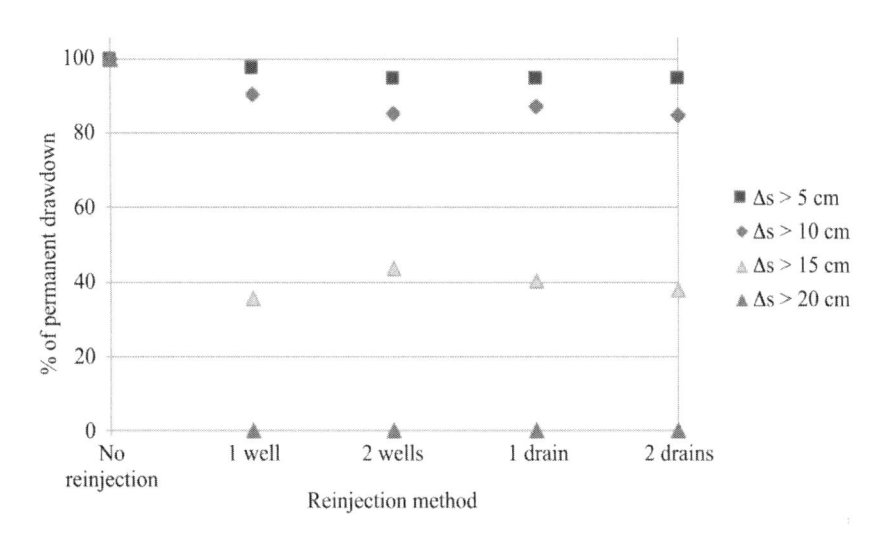

Figure 2.4. Percentage of permanent drawdown in the unconfined aquifer as a function of the method chosen to reinject water, shown for drawdowns greater than 5, 10, 15 and 20 cm. The results presented are the average number of runs as a percentage of the water reinjected for a full year of reinjection (four runs for each reinjection method tested and a reference run (no reinjection), 17 runs). For a color version of this figure, see www.iste.co.uk/salles/estuarinecities.zip

Δs	Indicator	Without compensation	One inject. well	Two inject. wells	One infilt. drain	Two infilt. drains
> 5 cm	km²	89	88	84	85	82
	% surface area of Δs	100	98	94	96	92
> 10 cm	km²	43	40	37	37	38
	% surface area of Δs	100	92	85	86	87
> 15 cm	km²	16	6	6	4	5
	% surface area of Δs	100	37	41	26	30
> 20 cm	km²	6	0	0	0	0
	% surface area of Δs	100	0	0	0	0

Table 2.2. Enduring drawdown (areas and percentages) in the unconfined aquifer, depending on the method chosen to reinject water, shown for drawdowns over 5, 10, 15 and 20 cm. The model was run with 5% of the extracted water reinjected for one year, starting on day 1 of extraction

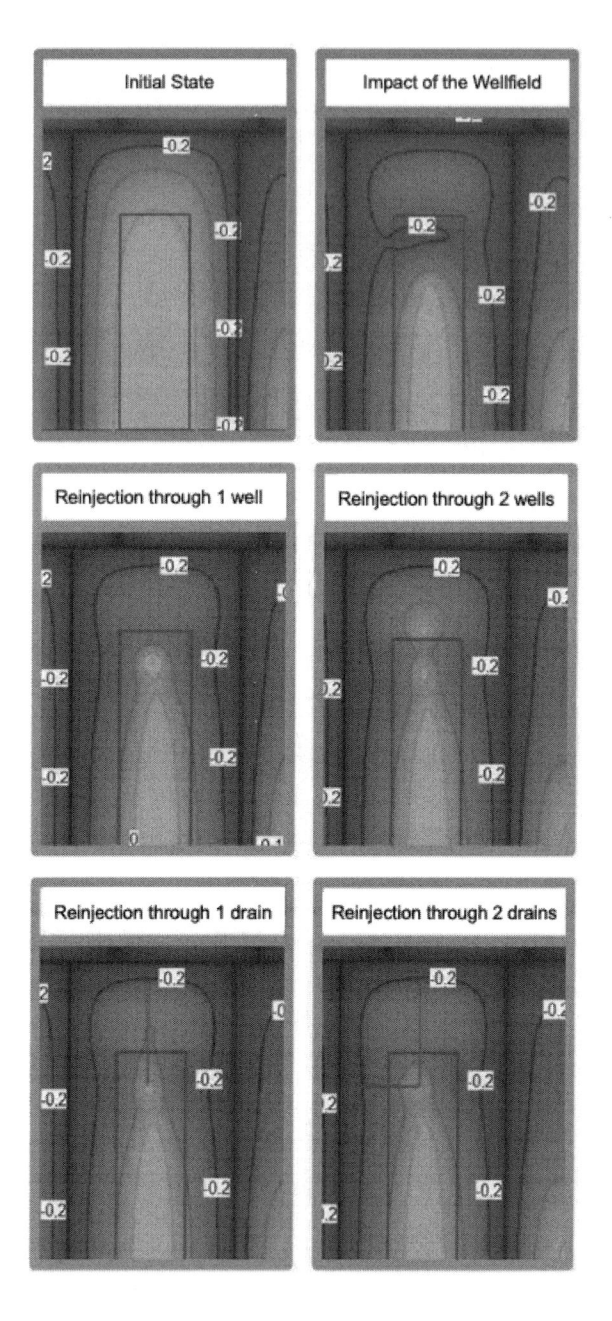

Figure 2.5. *Piezometric maps showing the free aquifer simulated by different model configurations. For a color version of this figure, see www.iste.co.uk/salles/estuarinecities.zip*

In order to compare the results of each reinjection method, the model is run with 5% of the extracted water reinjected for one year, starting on day 1 of the extraction. The results are presented as piezometric maps (Figure 2.4) and tables of remaining drawdowns after compensation (Table 2.2). The initial state is the model operating without water extraction or reinjection. The wellfield impact is the model operating for one year with the wellfield extracting water (10 mm^3/year). In the case of reinjection through the well(s), the water head, the piezometric level of -0.2 m in the middle, is pushed to the sides of the wellfield. Local flooding occurs around the reinjection site. In the case of reinjection through the drain(s), the water head, the piezometric level of -0.2 m in the middle, is also pushed to the sides of the wellfield. No local flooding is observed. In conclusion, spatialized reinjection is preferable to a single recharge well or ditch to avoid local flooding and maximize the spatial effectiveness of mitigation.

2.4. When and where should the extracted water be reinjected?

Based on the modeling, the percentage of extracted water reinjected must be sufficiently low so as not to interfere with the operation. The optimal value of 5% found from the numerical simulations seems to be very reasonable in this respect.

In the case of well reinjection, the percentage of water reinjected above a threshold produces local saturation and local flooding, rather than an increase in the area where mitigation is effective. In the case of recharge ditches, increasing the percentage of water reinjected moderately but consistently, improves the effectiveness of remediation. The optimal percentage must therefore be selected by balancing the benefit (effectiveness of the remediation) with the cost (a smaller volume of water available for domestic supply).

The timing of the start of reinjection is important for management. It should be further investigated using a high-resolution geological and hydrogeological model. One recommendation is to monitor the impacts of extraction and only begin the mitigation plan when negative impacts are observed. However, the simulation suggests that a delay in mitigation action significantly reduces its effectiveness.

A small difference in the size of the area where mitigation is effective is observed between reinjection through wells or recharge ditches. Therefore, since a significant number of shallow wells and surface ditches already exist in the region, the choice should be made based on the existing hydraulic infrastructure present at the future wellfield site. This decision must also consider the need to protect areas, primarily around the wellfield.

In addition to the potential of the reinjection solution, drainage management can be another option to consider. The wellfield area is a former marshland that was reclaimed in the 19th century by numerous deep drainage ditches to remove excess water in winter. One option would be to replace these ditches with flatter channels that could both drain excess groundwater and allow for the gradual recharging of the unconfined aquifer during dry periods (Figure 2.6).

Figure 2.6. *A wide rather than deep drainage ditch to avoid draining wetlands important for biodiversity at the Captieux firing site, a former NATO base and military training site, today a listed natural site for its rich biodiversity (source: Biteau 2018). For a color version of this figure, see www.iste.co.uk/salles/estuarinecities.zip*

The first experiments carried out make it possible to conserve and mobilize a large quantity of water that is usually discharged by a classic drainage system each year. According to a first simplified model, this volume could reach up to 25 mm^3/year. This experience shows that the water volumes in drainage ditches can be a significant water resource reserve.

A network of approximately 15 flow stations is being studied, which should soon provide a more realistic estimate of these potential reserves.

A potential obstacle to this ecological engineering technique is the differentiated land status of the drainage networks, which alternate between public, private and military domains within the region. The connection to the forest fire protection network is also a decisive factor.

2.5. Conclusion

Faced with the uncertainties around the impact of water abstraction from the deep aquifer on forestry activities, which are fueling the controversy in the Landes du Médoc catchment area, an ecological engineering technique was investigated through modeling. According to the numerical model used, an effective mitigation technique would be a partial reinjection of the water into the unconfined aquifer. A small part of the water extracted from a deep aquifer could be used to compensate, at least in part, for the negative impact of this water extraction on the shallow aquifer. The volume of water required to achieve a significant mitigation effect, potentially sufficient to avoid negative impacts on vegetation, is estimated to be 5%, which is low. Immediate initiation of the reinjection process is an important factor in maximizing mitigation effectiveness. Modifying drainage techniques traditionally associated with silvicultural activity from deep ditches to wide drains could slow down runoff and gradually recharge the open water table. These ecological engineering techniques, which are still at the experimental stage, could be included in the next consultations planned by the project owner, Bordeaux Métropole, with all the stakeholders.

2.6. References

AcclimaTerra, Le Treut, H. (ed.) (2020). Anticipating climate change in Nouvelle-Aquitaine. To guide policy at local level [Online]. Available at: https://www.acclimaterra.fr/wp-content/uploads/Synthese-AcclimaTerra-EN-1.pdf?utm_source=site+web&utm_medium=click&utm_campaign=2023.

Barthel, R., Ekström, L., Ljungkvist, A., Granberg, M., Merisalu, J., Pokorny, S., Banzhaf, S. (2017). Combining scientific and societal challenges: A water supply case study from the Koster Islands, Sweden. *EGU General Assembly Conference Abstracts*, 19, 9294.

Biteau, B. (2018). Forum – La lettre des marais atlantiques. *Forum des marais atlantiques*, 36, ISSN 1769–0013.

Blair, P. and Buytaert, W. (2016). Socio-hydrological modelling: A review asking "what and how?" *Hydrology and Earth System Sciences*, 20, 443–478.

Bordeaux, M. (2015). Projet de champ captant des Landes du Médoc. Étude des relations eaux souterraines – Eaux superficielles. Report, A80589/C, Naldeo, Antea Group, Bordeaux.

BRGM (1997). Schéma Directeur de gestion de la ressource en eau du département de la Gironde. Simulation d'un champ captant la nappe de l'Oligocène dans le secteur de Sainte-Hélène à l'aide d'un modèle gigogne couplé au modèle nord-aquitain. Report, R 39684, 71, Orléans.

Dillon, P. (2005). Future management of aquifer recharge. *Hydrogeology Journal*, 13, 313–316.

Doble, R.C. and Crosbie, R.S. (2016). Review: Current and emerging methods for catchment-scale modelling of recharge and evapotranspiration from shallow groundwater. *Hydrogeology Journal*. doi: 10.1007/s10040-016-1470-3.

EGID – Université Bordeaux Montaigne (2001). Établissement des conditions de gisement (limites et exutoires potentiels) des aquifères tertiaires du Médoc, à partir des données existantes. Report, EGID, Université Bordeaux Montaigne, Bordeaux.

EGID – Université Bordeaux Montaigne (2003). Identification et quantification des ressources des aquifères tertiaires du Médoc (Miocène, Oligocène, Éocène) à l'échelle du 1/100 000. Report, EGID, Université Bordeaux Montaigne, Bordeaux.

Grau-Martinez, A., Floch, A., Torrento, C., Valhondo, C., Barba, C., Domenech, C., Soler, A., Otero, N. (2018). Monitoring induced denitrification during managed aquifer recharge in an infiltration pond. *Journal of Hydrology*, 561, 123–135.

Jollivet, M. and Legay, J.-M. (2005). Dossier Interdisciplinarité Canevas pour une réflexion sur interdisciplinarité entre sciences de la nature et sciences sociales. *Natures Sciences Sociétés*, 13(2), 184–188.

Loustau, D., Picart, D., Saltel, M. (2016). Impact de l'exploitation d'un champ captant de 10 Mm^3/an dans la nappe de l'Oligocène sur le fonctionnement des peuplements de Pin maritime du Sud Médoc. Oral presentation.

Massuel, S., Perrin, J., Mascre, C., Mohamed, W., Boisson, A., Ahmed, S. (2014). Managed aquifer recharge in South India: What to expect from small percolation tanks in hard rock? *Journal of Hydrology*, 512, 157–167.

Massuel, S., Riaux, J., Molle, F., Kuper, M., Ogilvie, A., Collard, A-L., Leduc, C., Barreteau, O. (2018). Inspiring a broader socio-hydrological negotiation approach with interdisciplinary field-based experience. *Water Resources Research*, 54, 2510–2522.

Murray, R., Louw, D., van der Merwe, B., Peters, I. (2018). Windhoek, Namibia: From conceptualising to operating and expanding a MAR scheme in a fractured quartzite aquifer for the city's water security. *Sustainable Water Resources Management*, 4, 217–223.

Pédron, N., Abou Akar, A., Gomez, E. (2009). Simulation d'impact d'un champ captant dans l'aquifère Oligocène sur le secteur de Ste-Hélène (33) à l'aide du Modèle Nord-Aquitain (MONA). Report, BRGM/RC-57035-FR, Orléans.

Riaux, J. and Massuel, S. (2014). Construire un regard sociohydrologique (2). Le terrain en commun, générateur de convergences scientifiques. *Natures, sciences, sociétés*, 22, 329–339.

Roux, J.C. (2006). Bassin Aquitain. In *Aquifère et eaux souterraines en France*, volume 1, Roux, J.C. (ed.). BRGM, Paris.

Saltel, M. and Arnaud, L. (2015). Modèle phonème : construction, paramétrisation et évaluation qualitative et statistique du calage en régime transitoire. Report, BRGM/RP-65368-FR, Orléans.

Saltel, M., Picart, D., Loustau, D., Pédron, N. (2016). Multi-model approach to evaluate the impact of a future well field on forest production (South-West of France). *Proceedings of the 43rd IAH Congress*. Montpellier.

Sivapalan, M. and Blosch, G. (2015). Time scale interactions and the coevolution of humans and water. *Water Resources Research*, 51(9), 6988–7022.

Sivapalan, M., Savenije, H.H.G., Blöschl, G. (2011). Socio-hydrology: A new science of people and water. *Hydrological Processes*. doi: 10.1002/hyp.8426.

Vigneaux, M. (1975). *Guides géologiques régionaux : Aquitaine occidentale*. Masson, Paris.

Vincent, A. (2007). Étude hydrologique et hydrogéologique du bassin sédimentaire côtier de Kaluvelli-Pondichéry, Tamil Nadu, Inde. PhD Thesis, Université Pierre et Marie Curie, Paris.

Wesselink, A., Kooy, M., Warner, J. (2016). Socio-hydrology and hydrosocial analysis: Towards dialogues across disciplines. *WIREs Water*, 4, e1196 [Online]. Available at: https://doi.org/10.1002/wat2.1196.

Westerberg, I.K., Di Baldassarre, G., Beven, K.J., Coxon, G., Krueger, T. (2017). Perceptual models of uncertainty for socio-hydrological systems: A flood risk change example. *Hydrological Sciences Journal*, 62(11), 1705–1713.

PART 2

Protecting Against Risks, by the Estuary, and for the Estuary

3

Living in a City Exposed to Flood Risk: At What Cost(s)?

Climate change is putting large estuarine and coastal cities at increasing risk of flooding, with an increase in the frequency and intensity of storms and floods (IPCC 2014, 2019, 2021). At the same time, it is estimated that nearly 65% of the world's population will live in urban areas by 2050, implying a new allocation of resources and services to the population (United Nations 2019). In France, the estimated population in 2050 is expected to be around 74 million inhabitants, distributed mostly around urban centers, particularly in the west and south (Desrivierre 2017), increasing the vulnerability of coastal cities to global change.

In this medium- and long-term perspective, cities must address new governance challenges to meet the imperative of implementing climate change adaptation measures. These measures must allow for sustainable development, ensuring quality of life and safety for their inhabitants and preserving biodiversity (Dejean et al. 2019). The GEMAPI[1] regulation (management of aquatic environments and flood prevention) is part of the measures that are imposed on metropolises (in the wake of the 2014 MATPAM "law on the modernization of territorial public action and the affirmation of metropolises", and the 2015 NOTRe "law on the new territorial organization of the Republic") even though they are confronted with ever-increasing residential attractiveness (November 1994). The issue of sustainable urban development, combined with that of flood risk

---

Chapter written by Jeanne DACHARY-BERNARD and Florian VERGNEAU.
1 *Gestion des Milieux Aquatiques et Prévention des Inondations.*

management, leads decision-makers to question the development choices to be made. Understanding the way people live in the territory (particularly what type of housing they live in, where, since when) allows us to identify possible levers for public action likely to influence future residential locations in the context of global changes.

Among the main flood risk management mechanisms, the flood risk prevention plan, PPRI[2], is a tool defined by regulation at the territorial level, aimed at controlling urbanization in flood-prone areas. The objective is to control urban development through more or less restrictive zoning (Douvinet et al. 2011). Alongside this, there is another national system, of insurance: the so-called CatNat ("*catastrophe naturelle*": "natural disaster") system. CatNat makes it possible to trigger compensation for damage from public funds in the event the public authorities declare a natural disaster (Barraqué 2014). These mechanisms – preventive and curative – can both be mobilized at the scale of territories exposed to flood risk, hence our hypothesis that they impact residential processes.

In this context, the study mobilized here explores the relationship between flood risk and residential choices through an analysis of real estate prices. To what extent are the prices of real estate transactions that have taken place in a territory impacted by the flood risk? More precisely, how do risk management strategies influence prices? Do preventive and curative mechanisms have a cumulative effect on these residential logics? In this chapter, we propose to analyze the risk/amenity trade-offs made by property buyers, by observing house prices, and in particular the complexity of the effect of risk on prices according to the risk management mechanisms implemented. We seek to highlight the ways in which the risk of flooding is or is not integrated into the real estate market, in particular by distinguishing between the preventive and restorative mechanisms associated with the risk of flooding; Bordeaux, an estuarine city subject to the risk of flooding-submersion, is an ideal case to try to answer these questions.

We first discuss how economics explains residential choices, particularly with respect to environmental factors. We then introduce the field of study, that is, Bordeaux city and its neighboring estuarine municipalities, as well as the data used. Finally, we present the modeling carried out to understand what explains real estate prices. Some concluding elements provide information on the main processes that guide property pricing in the

---

2 *Plan de Prévention du Risque Inondation.*

Bordeaux metropolitan area. In particular, we emphasize the simultaneous presence of peri-urbanization around the city and rurbanization of the more remote areas. We also show that exposure to flooding risk infers a discount in terms of the real estate market value, although this differs according to the capacity of municipalities to respond to the need for damage compensation.

3.1. Residential location and risk as economic issues

Urban economics has long been interested in the determinants of households' residential choices: why people live in one place and not another. Initially based on the reference framework of monocentric North American cities (Alonso 1964; Mills 1967; Muth 1969), this literature has historically shown the importance of accessibility to downtown jobs in the location choices of households under income constraints. In order to explain new spatial configurations of territories, in response to the processes of peri-urbanization in particular (Cavailhès et al. 2003) or "coastalization" (Dachary-Bernard et al. 2011), economics has progressively taken into account other factors, such as amenities, which are considered geographically situated characteristics that provide well-being to people (Bartik and Smith 1987). Therefore, peri-urbanization is explained by the desire of households to benefit from green amenities with gardens and a lower density of buildings in the neighborhood. Amenities related to proximity to the sea (sea views, access to swimming areas, etc.) also explain the process of coastal transformation that has been underway for several decades. Such amenities may be endogenous (i.e. produced by the city itself) or exogenous (pre-existing in the city, such as a remarkable space) (Brueckner et al. 1999). These are central to accounting for current urban developments. Conversely, some externalities can generate negative effects, such as noise pollution in the vicinity of a freeway or visual pollution near industrial or commercial areas. Such nuisances have a dissuasive effect on residential choices (Travers et al. 2009). Some studies (Cavailhès et al. 2014) have concluded on the importance of climatic variables, temperature and rainfall on residential locations, showing that urban sprawl and global warming are mutually reinforcing in the sense of an ever-increasing whittling away of urban peripheral spaces. We may therefore wonder whether exposure to risk plays the role of a "desamenity", a dissuasive to residential location choices.

The economic mechanism at work to explain the real estate pricing is the capitalization of amenities in terms of price: the proximity or presence of amenities produces a "premium" that increases market property values. This premium refers to the logic of direct exposure or accessibility. A similar logic should therefore be able to be applied to the "desamenity" associated with the risk of flooding: it would lead to a loss in value of the properties directly exposed or close to the source of the risk.

Previous studies have shown that flood risk is capitalized in prices, but not always in the same way. A discount is often observed, with properties located in flood-prone areas being less expensive, all other things being equal, than properties located in non-vulnerable areas (Bin and Polasky 2004; Kousky 2010). In addition, some characteristics of flooding influence this capitalization, such as the frequency or severity of the event (Tobin and Montz 1989, 1990), or the repetition of a natural disaster, which can lower the price permanently (Tobin and Newton 1986). The temporal proximity to the disaster is an important factor. The over-occurrence of events acts as a source of updated risk information (Atreya et al. 2013, p. 578). However, some work has been able to show that the effect of risk can also be positive. The frequency of flooding can create a form of learning at the level of public actors, particularly on how to respond, and can lead them to renovate some public infrastructure at the local level, which would have a positive impact on prices (Tobin and Montz 1989). The experience of flooding can also create this learning effect at the individual level, motivating residents to adopt adaptive (Siegrist and Gutscher 2008) or preparedness behaviors (Becker et al. 2017) that can then be capitalized into prices. Finally, this capitalization of risk in prices can also be influenced by insurance premiums and the ability of the insurance system to be mobilized to reduce risk (Surminski et al. 2015).

However, real estate in vulnerable areas continues to command very high prices, raising doubts about the actual integration of risk. This would be explained by the amenities also present in these vulnerable areas, such as sea views, accessibility to recreational activities, the value of which would largely exceed the discount associated with the risky nature of these settlements. The literature has also appropriated this topic of the amenity/risk trade-off (Bin et al. 2008), showing the predominance of positive effects over negative aspects of proximity to water. Mauroux (2015a) points out that French case studies (Longuépée and Zuindeau 2001; Hubert et al. 2003) have highlighted a negative capitalization of risk in terms of prices that are often counterbalanced by the positive effect of "amenities". The perception of flood risk held by individuals, as well as by the insurance mechanisms at work, seems to be an important factor in these inefficiencies (Mauroux 2015a).

3.2. Empirical strategy of the hedonic price model

The hedonic price model (HPM), originally developed by Rosen (1974), has been widely used to study real estate and land pricing. It is a method based on Lancaster's (1966) consumer theory, according to which the value of a good is made by its various characteristics. In other words, this means that property price can be explained by different variables such as structural or "intrinsic" characteristics, neighborhood characteristics and environmental characteristics. (Freeman III 1979).

We sought to explain the impact of different variables on house prices by estimating a hedonic price function. There is no consensus on how best to formalize this hedonic function (Dubé et al. 2011). We opt for a mixed-form model; as is common in the literature, the price is expressed in logarithmic form and the explanatory variables of the model are expressed in linear or logarithmic form (areas, distances, densities) according to the nature of these variables.

Therefore, the marginal effect of a one-unit change in an explanatory variable on price will be measured as a percentage or will be directly interpretable. If explanatory variables are also expressed in logarithmic form, then the estimated coefficient can be interpreted directly as a price elasticity. This semi-logarithmic form is consistent with Rosen (1974) and has been widely validated in the literature (Bourassa et al. 2004). Hedonic price models distinguish between three sets of attributes that can affect real estate values: "housing attributes" or intrinsic characteristics, "neighborhood characteristics" where the housing is located, which refer to a demand for social identification, and "accessibility variables" that satisfy a demand for integration into the market for labor, goods, leisure, etc. (Baumont and Legros 2013). We add to these standard characteristics the risk[3] variables assumed to reflect the demand for protection from risk.

---

3 The hedonic model used is therefore the following:

$$\log P = \alpha + \sum_{k} \beta_k Z_k + \sum_{g} \beta_g E_g + \sum_{h} \beta_h L_h + \sum_{m} \beta_m N_m + \sum_{n} \beta_n V_n + \varepsilon$$

where log P is the vector of the price in logarithmic form of the goods; Z, E, L, N and V are the vectors of the intrinsic, environmental, location, neighborhood and risk characteristics of the goods, respectively and ε represents the vector of assumed iid error terms.

Moreover, real estate is by nature spatial, which means that we must take into account possible spatial effects, that is, the prices of properties observed in one place are not independent of properties in their neighborhood. The Moran test that we have implemented does show the existence of a positive spatial dependence between the prices of the properties studied (indicating a relative concentration of similar observations), but tells us nothing about the type of spatial effect to correct. To this end, we define, by means of a spatial weight matrix, the form (neighborhood or distance) of the relationship between properties that best reflects the processes at work. In the case of our study, the weight matrix best suited to our data turns out to be the one-nearest neighbor matrix. Based on this form of spatial interdependence, several econometric models were estimated in order to identify how different factors affect housing prices. Based on the results of Lagrange tests and the common factor[4], we retain the Spatial Durbin Model (SDM) (Halleck Vega and Elhorst 2015), whose results are presented later.

3.3. Bordeaux Métropole study area and data

The Bordeaux metropolitan area is subject to a fluvio-maritime regime, under the influence of both the Garonne and Dordogne rivers and the ocean. Eighteen municipalities in the Bordeaux metropolitan area are thus exposed, in whole or in part, to the risk of flooding. One-third of the territory is located under the high waters of the Garonne, and more than 40,000 people live in flood-prone areas. Since January 1, 2016, the Bordeaux Métropole, through GEMAPI, has assumed full responsibility for the management of aquatic environments and flood prevention. As part of the implementation of the "flooding" directive, a local flood risk management strategy, TRI[5] (2016-2021) has been defined and implemented on the scale of territories at significant risk of flooding by Bordeaux Métropole. One of the axes of this strategy concerns the reduction in the vulnerability of goods and people. Assets and activities in flood-prone areas have been identified and pose numerous challenges in economic, agricultural, natural and human terms. The associated damage for an average event in Bordeaux Métropole has been estimated at approximately 190 million euros.

---

4 We have chosen not to present these tests so as not to be weighed down by the technical aspects of the subject.
5 *Territoires à Risque important d'Inondation.*

However, other Gironde estuarine municipalities are also exposed to the risk, and the reflections on flood risk management are expanding from the metropolitan perimeter to the estuary scale, with the constitution of a flood prevention action program, PAPI[6], concerning the period 2016–2021. In order to study the metropolitan real estate market in terms of the flood risk, which has an impact beyond the metropolitan perimeter, our field of study concerns Bordeaux city and its neighboring estuary territories along the river, which are also subject to the flood risk, as shown in Figure 3.1a.

In order to study real estate transactions (especially houses, see Figure 3.1b) over the period 2012–2016, we mobilized the requests for land value (DVF in French[7]) database from the French treasury department (DGFiP[8]), which lists real estate transactions from notarized deeds and cadastral information. We also mobilized the geographical databases "Topo", "Address" and "BD Plot of land" from the IGN[9], as well as socio-economic data produced by the French national institute of statistics and economic studies (INSEE[10]). The data on flood risk were taken from the national GASPAR[11] database of the general directorate for risk prevention by the "Ministry of Ecological and Transition Solidarity"[12]. The variables used to conduct the analysis are listed in Table 3.1.

We have selected 10 common intrinsic attributes to characterize houses, for which we expect to observe a typical effect on prices. Temporal variables are included to capture pure real estate dynamics.

The residential environment of the transactions is described by nine synthetic variables that capture processes of accessibility (distance variables), proximity (share of vegetation, density and average height of the buildings) and social environment of municipalities (social composition of neighborhoods and enclaves index).

---

6 *Programme d'actions de prévention des inondations.*
7 *Demandes de Valeur Foncière.*
8 *Direction Générale des Finances Publiques.*
9 *Institut national de l'information géographique et forestière.*
10 *Institut national de la statistique et des études économiques.*
11 *Gestion assistée des procédures administratives relatives aux risques naturels.*
12 *Le ministère de la Transition écologique et solidaire.*

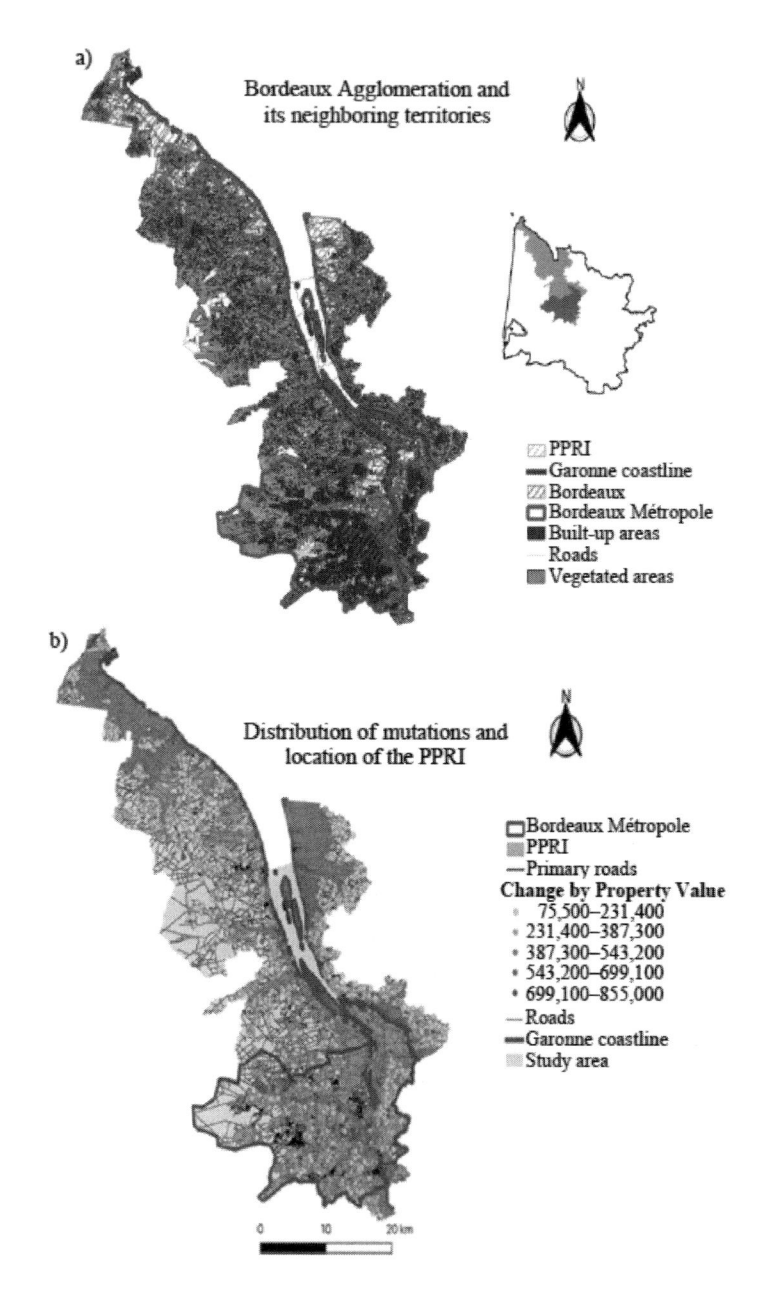

Figure 3.1. *a) The Bordeaux city and its neighboring territories. b) location of house mutations according to the PPRI. For a color version of this figure, see www.iste.co.uk/salles/estuarinecities.zip*

Variable	Description	Mean/distribution	Std. Dev.	Min.	Max.
	Intrinsic variables				
transact_price	Price of the transaction in Euros	268,887.6	129,928	75,550	855,000
mut_year 201X	Year of the mutation	2011: 4,391 2012: 4,078 2013: 4,196 2014: 3,690 2015: 4,220 2016: 4,296	–	–	–
surf_land	Land area in m²	762.8	5,188.6	0	625,596
nb_mr	Number of main rooms	4.2	1.3	1	14
surf_build	Surface of the building in m²	101.7	39.3	10	405
terrace	Presence of a terrace	0 = 19,617 1 = 5,254	–	–	–
swim_pool	Presence of a swimming pool	0 = 22,719 1 = 2,152	–	–	–
bath_more	More than one bathroom	0 = 18,953 1 = 5,918	–	–	–
garage	Presence of a garage	0 = 9,673 1 = 15,198	–	–	–
good_cond	House in good condition	0 = 4,439 1 = 20,432	–	–	–

Variable	Description	Mean/distribution	Std. Dev.	Min.	Max.
Intrinsic variables					
dist_BX	Distance to Bordeaux City Hall in meters	12,738.0	16,944.6	147.9	89,723.1
dist_road	Distance to nearest primary road in meters	3,654	8,229.9	13.7	47,646.1
dist_river	Distance to the Garonne in meters	4,583.8	3,678.3	19.5	19,926.1
metrop	Transaction realized within Bordeaux Métropole	0 = 5,027 1 = 19,844	–	–	–
surf_nature	Share of vegetation within a radius of 200 meters in m²	22,817.7	20,289.3	0	11,6784.2
surf_build_n	Density of buildings within 200 meters in m²	7,843.8	5,506.3	15	38,514.7
height_build_n	Average height of buildings within 200 meters in meters	6.71	0.9	1.6	9.2
comp_socialeX	Social composition of the neighborhood at the IRIS level	G1: 8,796 G2: 2,226 G3: 13,849	–	–	–
ind_enclav	Enclosure indicator approaching in a multivariate way the equipment rate of the IRIS	8,975.3	4,495.9	635.7	20,705.6
Flooding Variables					
PPRI	Mutation realized or not in a PPRI zone	0 = 23,043 1 = 1,828	–	–	–
CatNat	Mutation within a municipality that has experienced at least nine natural flood-related disasters	0 = 8,002 1 = 16,869	–	–	–

Table 3.1. *Variables used in the model and their main characteristics*

Finally, two variables are used to characterize the flood risk. The first is constructed from the geographical layer of regulatory zonings of the flood risk prevention plans (PPRI) from the "Géorisque" database. It is a binary variable that indicates whether or not the mutation is located in a PPRI zone (after checking that the PPRI has been in place before the transaction). This first variable is part of a preventive approach to flood risk management, in the sense that this zoning provides information about exposure to the risk and the associated urban planning constraints. A second variable captures a remedial dimension associated with flooding. In this more curative dimension of risk management, the State may, in response to the municipality's request, recognize the flooding event as a natural disaster (CatNat mechanism), thus giving the right to compensation for damages by insurers. The constructed variable provides information on the intensity of the reactivity of the communities, on their capacity to trigger compensation; it is defined in relation to the threshold of nine decrees, and thus translates the dynamics at the local level of the CatNat repair mechanism.

We hypothesize at this point that the preventive and reparation aspects associated with flood risk impacts house prices.

3.4. A multifaceted city

We present and interpret the results in two stages. First, we highlight the main trends that impact pricing, and then we focus on the specific influence of the variables of interest, that is, those related to flood risk[13].

	$\hat{\beta}$	$\hat{\theta}$
constant	7.754 (0.1346)	—
surf_build	0.005 (0.001)***	-0.0002 (0.0001)***
surf_land	0.046 (0.0016)***	-0.011 (0.0017)***
nb_mr	0.012 (0.0021)***	-0.003 (0.0021)***
terrace	0.070 (0.0046)***	0.012 (0.0046)***
swim_pool	0.178 (0.0068)***	-0.014 (0.007)***
bath_more	0.070 (0.0049)***	0.003 (0.005)***

---

13 The results of the model are presented in Tables 3.2 and 3.3. They are presented in the form of the coefficients estimated by the hedonic model (Table 3.2) and in the form of the direct or indirect effects of the variables on prices (Table 3.3).

	$\hat{\beta}$	$\hat{\theta}$
garage	0.036 (0.0041)***	-0.015 (0.0042)***
good_cond	0.061 (0.005)***	-0.004 (0.0051)***
dist_road	0.030 (0.099)*	0.075 (0.0991)*
ind_enclav	-0.012 (0.0807)*	0.098 (0.0808)*
dist_BX	< -0.000 (0.000)***	< -0.000 (0.000)***
dist_river	0.017 (0.0234)**	0.024 (0.0234)**
surf_nature	0.008 (0.0048)***	0.002 (0.0048)***
surf_build_n	-0.103 (0.0348)**	0.044 (0.0348)**
height_build_n	0.106 (0.0365)**	0.024 (0.0366)**
comp_social1	Ref	Ref
comp_social2	0.047 (0.034)**	-0.024 (0.034)**
comp_social3	0.063 (0.0279)**	0.061 (0.028)**
metrop	0.013 (0.1066)	0.152 (0.1067)
CatNat	0.058 (0.0493)**	0.006 (0.0493)**
PPRI	-0.074 (0.0234)**	-0.023 (0.0238)**
mut_year2011	Ref	Ref
mut_year2012	0.014 (0.0061)***	-0.011 (0.0061)***
mut_year2013	0.021 (0.0061)***	-0.012 (0.0061)***
mut_year 2014	0.016 (0.0063)***	-0.004 (0.0063)***
mut_year 2015	0.021 (0.006)***	-0.007 (0.0061)***
mut_year 2016	0.058 (0.006)***	-0.012 (0.0061)***
dist_road:ind_enclav	0.0004 (0.0109)**	-0.011 (0.0109)**
CatNat:PPRI	0.006 (0.0336)**	0.050 (0.0338)**
Rho (p)	0.161 (0.000)***	
Lambda (p)	–	
LR test (p)	1,035.3 (0.000)***	
z-value (p)	33.5 (<0.000)***	
Wald statistic (p)	1,119.1 (0.000)***	
Number of observations	24,871	
Moran I on residuals (p-value)	-0.008 (0.8098)	

Table 3.2. *Results of the SDM estimation*

COMMENT.– $\hat{\beta}$ and $\hat{\theta}$ are the estimated parameters of the variables expressed respectively in terms of their level and spatially lagged form. The risk levels of the tests are denoted by: 1% "***", 5% "**" and 10% "*". Standard deviations are presented in brackets.

3.4.1. Confirmed metropolitan trends coupled with emerging rurbanization

The model that best fits the data is a spatial hedonic model[6]. This means that there are spatial effects that are expressed through spatially lagged variables. That is, there is a diffusion of prices and characteristics of a property with neighboring properties. In other words, a characteristic of a property will impact the price of the property, as well as the price of properties in the vicinity. The indirect effects in Table 3.3 reflect this. For example, it can be seen that the price of a house will be higher if it has a swimming pool (positive direct effect), as well as if the properties in the neighborhood have a swimming pool (indirect effect). This means that residents' perceptions of their residential environment depend, in part, on the perceptions of their neighborhood and their home group (family/friends) (Peacock et al. 2005). This first result therefore indicates that prices in the city and its neighboring estuarine territories are part of a spatial logic and respond to a relationship of interdependence between territories.

3.4.1.1. Houses valued for their characteristics…

The results relating to the intrinsic characteristics of the property indicate, in line with the literature (Baumont and Legros 2013), that the price of a residence will be higher if it has a large built-up area, a large area of land, a large number of rooms, a terrace, a swimming pool, more than one bathroom, a garage, and if it is generally in good condition. In addition to these structural variables, the time variables are all found to have a positive effect on prices (relative to the base year 2011), and this effect increases over the period. This refers to the general price increase that the real estate market has experienced over the period in this area (A'urba 2021). Therefore, over the period 2012–2017, house prices increased by 12% in Gironde and 16% in Bordeaux Métropole, with an average price per square meter of €2,450 and an average value in 2017 of €268,000 in Gironde. For

---

6 In particular, the spatial Durbin model by Halleck Vega and Elhorst (2015).

comparison, our sample has an average price for houses in our study area (i.e. the extended city) over 2012–2016 that is quite similar at €268,888 (Table 3.1).

	Direct effect	Indirect effect	Total effect	MCO
surf_build	0.005***	0.0006***	0.005***	0.005 ***
surf_land	0.45***	-0.004***	0.041***	0.045 ***
nb_mr	0.012***	-0.001***	0.011***	0.012 ***
terrace	0.073***	0.025***	0.098***	0.077 ***
swim_pool	0.180***	0.016***	0.196**	0.186 ***
bath_more	0.072***	0.016***	0.088***	0.076 ***
garage	0.034***	-0.010***	0.024***	0.029 ***
good_cond	0.061***	0.006***	0.067***	0.061 ***
dist_road	0.038***	0.087*	0.125**	0.128 ***
ind_enclav	-0.003*	0.105*	0.102**	0.104 ***
dist_BX	0.000***	0.000***	0.000***	< -0.000 ***
dist_river	0.020**	0.029**	0.049***	0.047 ***
surf_nature	0.008***	0.004***	0.012***	0.013 ***
surf_build_n	-0.100**	0.030**	-0.070***	-0.076 ***
height_build_n	0.110**	0.044**	0.154***	0.149 ***
comp_social2	0.046**	-0.018**	0.028**	0.028 ***
comp_social3	0.071**	0.078***	0.149***	0.156 ***
metrop	0.029	0.168	0.197	0.195 ***
CatNat	0.059**	0.016**	0.075***	0.075***
PPRI	-0.078**	-0.038**	-0.116**	-0.114 ***
mut_year2012	0.013***	-0.010***	0.003***	0.009
mut_year2013	0.20***	-0.009***	0.011***	0.016 **
mut_year 2014	0.015***	-0.001***	0.014***	0.014 **
mut_year 2015	0.021***	-0.004***	0.017**	0.018 ***
mut_year 2016	0.058***	-0.002***	0.056***	0.055***
dist_road:ind_enclav	-0.001**	-0.011**	-0.012***	-0.013 ***
CatNat:PPRI	0.011**	0.056**	0.067***	0.058 ***

Table 3.3. *Direct, indirect and total effects of variables on prices. The risk levels of the tests are denoted as 1% "***", 5% "**" and 10% "*"*

3.4.1.2. ... But also for their location...

The results also indicate that the environment in which the house is located plays a role in its price through different dimensions.

Proximity to Bordeaux is an important element for buyers, with the negative coefficient meaning that as distance to the city center increases (i.e. as we move further away from Bordeaux), the price of the property will decrease. On the other hand, the low coefficient (almost zero) indicates that this gradient is only slightly decreasing, which suggests that the relationship to the city center is either captured by other variables or has little influence. The fact that the "city" variable is not significant in the model supports the second interpretation: accessibility to the central city (especially for its jobs) and membership of the metropolitan area, which theoretically have a strong influence on prices, do not seem to define a mechanism that structures prices within our study area. Our hypothesis that the logic of metropolitan attractiveness alone is not sufficient to explain price formation in our territory is thus confirmed.

Other factors that are typically observed to explain prices include access to transportation. Our model reveals a positive influence of distance to the main road on house prices. This means that houses are more expensive the farther they are from a major transportation route – all else being equal. Proximity to major transport infrastructure mainly induces nuisances that outweigh the accessibility benefits, as has been shown in the literature (Travers et al. 2009), except in the case of very remote housing for which access to the road network is perceived as an amenity.

We also observe a negative effect of proximity to the river on house prices (with positive coefficients of distance to the river). At the scale concerned by our study, it is therefore rather the blue disamenities that seem to play on prices rather than the recreational or landscape aspects.

3.4.1.3. ... Or for their natural and social environment

The influence of the property's immediate "natural" environment (i.e. a radius of 200 meters around the property) impacts prices through variables targeting the share of vegetation or the share of buildings. Therefore, a large vegetated area near the property drives prices upwards (as well as those of neighboring houses), reflecting the value of green spaces. On the other hand, an environment close to a built-up environment is somewhat negatively perceived, with a preference for a densely built neighborhood rather than one that is spread out. It should be noted that the indirect effect of the built-up area within 200 meters is positive, underlining the ricochet effect that this variable has: the fact that a residence is negatively impacted by a built-up environment close to it in turn positively affects the residence that does not have a built-up environment close to it; this is, in a way, an indirect enhancement.

The social environment also has an impact on house prices. The social composition of the neighborhood[7] has a positive direct effect on price (Table 3.3), reflecting the search for a socio-economically homogeneous neighborhood, and especially not for a mixed neighborhood. Here again, the indirect effect of this variable on price is negative, underlining the fact that a disadvantaged environment is not valued if it extends spatially. The enclavement index, defined on the basis of distance to main facilities, has a positive overall impact on prices, highlighting the value attributed to being isolated. By separating the direct and indirect effects (Table 3.3), we understand that the isolation of a residence is poorly perceived, but that the isolation from its neighbors positively affects its value. It should be noted that the cross-tabulation of this isolation index with the distance to the road network is negative, meaning that a very isolated house far from the main transport infrastructure, all other things being equal, is less expensive; as indicated above, proximity to the transport infrastructure is only valued in cases where it is isolated from basic facilities.

These results highlight the presence of the traditional process of peri-urbanization, motivated by the search for accessibility to the city center and to green amenities in the immediate vicinity of the residence. On the other hand, proximity to the main transport infrastructures mainly induces nuisances, except in the case of very isolated housing. The phenomenon of socio-spatial segregation, often highlighted in contemporary urban forms and in the case of the Bordeaux metropolitan area (Décamps and Gaschet 2013), seems to be confirmed here with the search for homogeneous and unmixed socio-economic environments (Baumont and Legros 2013). However, these results also raise questions, at this broader metropolitan scale, about the

---

7 This variable summarizes a certain amount of economic information at the IRIS level, based on INSEE data on employment, education and housing. It is a variable defined in three classes according to whether the socio-economic conditions of the IRIS are favorable (class 3), unfavorable (class 2) or mixed (class 1).

simultaneous emergence of a process of rurbanization through which spaces far from the city (re)become attractive (Bailly and Bourdeau-Lepage 2011). In fact, there seems to be a premium for being far from equipment and transport infrastructures, and an emphasis on the search for green spaces.

3.4.2. *A double effect of flood risk on prices*

In order to study how flood risk impacts prices, we used two separate variables: one referring to PPRI zoning; the other relating to the intensity of CatNat events at the municipal level. We assume a negative effect for these variables, which would translate into a discount for a residence exposed to flood risk. In reality, we observe that these two variables have a very significant effect on house prices, albeit in opposing directions. PPRI zoning does have a *negative* influence, in line with the literature (Bin and Kruse 2006). Property located in a flood zone is less expensive, all else being equal, than those located outside this zone. On the other hand, we find that the CatNat variable has a *positive* effect on price. Being located in municipalities that have experienced numerous floods classified by the State as *natural disasters* leads to a higher price, all other things being equal. This result, surprising at first glance, must be interpreted in relation to the curative (or post-event) dimension that the variable brings. Indeed, whereas zoning is a preventive device aimed at regulating land use in areas exposed to hazards, the CatNat declaration is a tool that can be mobilized by municipalities following an extreme event in order to facilitate damage compensation. This system therefore has a restorative character associated with a financial dimension, since it is a question of opening up damage compensation rights.

These results highlight two mechanisms at work simultaneously (which are self-perpetuating). The first, linked to the PPRI variable, refers to risk aversion, which means that the individual tends to move away from the risk and does not choose to settle in a flood zone. Of course, this assumes that individuals are aware of the risk because, as the literature has shown, information about the exposure to risk of a home must be known by buyers in the hope that the preventive measure will be effective (Mauroux 2015b). However, in our case, the transaction data all post-date the IAL[14] (buyer–tenant database), which since 2006 has obliged buyers or tenants of the

---

14 *Information Acquéreurs-Locataires.*

property to be informed of the natural risks incurred by the property in question. We can therefore reasonably consider that our data does not contain any information asymmetry bias (Akerloff 1970) with respect to this risk, and admit that the location of a property in a PPRI zone is known to the market.

The second mechanism linked to the effect of the CatNat variable is more complex and original and deserves more detailed explanation. As mentioned above, when a flood is declared by decree to be a natural disaster, this opens up the right to reparation through damage compensation by the State. Such compensation, as in any insurance system, limits the financial loss associated with the event for the owner. In this respect, our results highlight the fact that being located in a municipality that has experienced a significant number of events of this type generates a sort of "premium". For a buyer, the risk of financial loss (in case of material damage to their property following a flood) is minimized if they take into account that they will receive compensation. Therefore, when the local authorities have been reactive with regard to reparation procedures following previous floods, this can constitute a form of "precedent" and produce a form of insurance inferring that the buyer will be compensated if a new flood should occur.

This result can be put into perspective with the work concerning the factors and institutional drivers of adaptation to climate change (Adger 2000). Indeed, the capacity of a local institution to take measures that reduce the risk of the negative effects of flooding can be related to its capacity to adapt to flooding events (Næss et al. 2005). This positive effect of the CatNat variable can thus be likened to a premium for the "compensation capacity" of communities, which some authors have described as "financial security" produced by the CatNat system at the cost of a reduction in the effectiveness of prevention devices (Cazaux et al. 2019). Our results also show that the intersection of the two variables is significant and positive, showing that this "indemnifying capacity" premium is all the stronger the more the property is in a flood zone. This perverse effect has been analyzed in the literature as a crowding-out phenomenon, meaning that receiving State aid in the event of a disaster can decrease an individual's incentive to use private insurance (Kousky et al. 2018). Other studies that focus more specifically on the French insurance system have shown that the CatNat system does not create any particular incentive for individuals to take mitigation measures (Poussin et al. 2013). Note, however, that the sum of these two opposing effects is negative, reflecting a negative net effect on the price of a flood risk. The remedial effect of the CatNat mechanism thus produces a perverse effect in the sense that it strongly diminishes the expected effect of the PPRI preventive instrument, without however cancelling it out.

3.5. Conclusion

Our study questioned how flood risk is integrated into the real estate transactions that determine residential choices. If cities must today "promote new and emerging forms of anticipatory governance" (*author's translation*), as the URBEST project puts it, they must understand the way in which the socio-economic processes at work within the territory unfold and mobilize different forms of interdependence. Through the prism of real estate prices in the Bordeaux city, extended to its neighboring estuarine territories, we have shown how residential choices respond to issues related to climate change, and in particular to flood risk.

Our results highlight the importance of spatial mechanisms at work on residential location choices, which demonstrates the weight of socio-ecological interdependencies on the real estate market. These interdependencies are both spatial and institutional, as revealed by the double play of risk management mechanisms combining local and national scales, and curative and preventive dimensions.

The major trends at work in the Bordeaux Metropolitan area, extended to the estuary, highlight the simultaneous existence of peri-urbanization and rurbanization processes. Could this be an illustration or the beginnings of what some, such as Senil (2021), view as a "rural transition"? Residential choices are certainly still marked by a search for proximity to employment areas, but they also express a desire to distance themselves from urban amenities and show a valorization of rural areas. The latter seem to be sought after for their own sake, and not only for their accessibility to urban centers.

With regard to the issues related to flood risk, as we expected, the value of property is negatively impacted by its exposure to risk, and also by that of neighboring properties. On the other hand, the remedial mechanism represented by the CatNat system, which intervenes after an event recognized as a natural disaster, produces an effect contrary to the preventive effect of PPRI zoning. The capacity of a territory to deploy repair measures and, consequently, to allow for the indemnization of damage, is valued in the prices.

Would the condition for the emergence of effective adaptive governance with regard to flood risk not lie in better articulation of prevention and reparation tools? Should risk governance not deploy its intervention at the more local scale of the processes on which it must be able to act? On the other hand, when reading the still high real estate prices in certain areas exposed to risk, we can only wonder about the real consideration of risk in residential behavior. We may therefore wonder how such adaptive governance of risks in cities could truly involve citizens, so as to make them as responsible as local managers.

3.6. References

Adger, W.N. (2000). Institutional adaptation to environmental risk under the transition in Vietnam. *Annals of the Association of American Geographers*, 90(4), 738–758.

Alonso, W. (1964). *Location and Land Use. Toward a General Theory of Land Rent.* Harvard University Press, Cambridge.

Atreya, A., Ferreira, S., Kriesel, W. (2013). Forgetting the flood? An analysis of the flood risk discount over time. *Land Economics*, 89(4), 577–596.

A'urba (2021). Prix de l'immobilier résidentiel en Gironde et dans Bordeaux Métropole. Report, A'urba, Bordeaux.

Bailly, A. and Bourdeau-Lepage, L. (2011). Concilier désir de nature et préservation de l'environnement : vers une urbanisation durable en France. *Géographie, économie, société*, 13(1), 27–43.

Barraqué, B. (2014). The common property issue in flood control through land use in France. *Journal of Flood Risk Management*, 10(2), 182–194.

Bartik, T.J. and Smith, V.K. (1987). Urban amenities and public policy. *Handbook of Regional and Urban Economics*, 2, 1207–1254.

Baumont, C. and Legros, D. (2013). Nature et impacts des effets spatiaux sur les valeurs immobilières. *Revue économique*, 64(5), 911–950.

Becker, J.S., Paton, D., Johnston, D.M., Ronan, K.R., McClure, J. (2017). The role of prior experience in informing and motivating earthquake preparedness. *International Journal of Disaster Risk Reduction*, 22, 179–193.

Bin, O. and Kruse, J. (2006). Real estate market response to coastal flood hazards. *Natural Hazards Review*, 7(4), 137–144.

Bin, O. and Polasky, S. (2004). Effects of flood hazards on property values: Evidence before and after hurricane Floyd. *Land Economics*, 80(4), 490–500.

Bin, O., Crawford, T.W., Kruse, J.B., Landry, C.E. (2008). Viewscapes and flood risk: Coastal housing market response to amenities and risk. *Land Economics*, 84(3), 434–448.

Bourassa, S.C., Hoesli, M., Sun, J. (2004). What's in a view? *Environment and Planning A*, 36(8), 1427–1450.

Brueckner, J.K., Thisse, J.-F., Zenou, Y. (1999). Why is central Paris rich and downtown Detroit poor? An amenity-based theory. *European Economic Review*, 43(1), 91–107.

Cavailhès, J., Peeters, D., Sekeris, E., Thisse, J.-F. (2003). La ville périurbaine. *Revue économique*, 54(1), 5–23.

Cavailhès, J., Joly, D., Hilal, M., Brossard, T., Wavresky, P. (2014). Économie urbaine et comportement du consommateur face au climat. Effet sur les prix hédonistes et sur l'étalement urbain. *Revue économique*, 65(4), 591–619.

Cazaux, E., Meur-Ferec, C., Peinturier, C. (2019). Le régime d'assurance des catastrophes naturelles à l'épreuve des risques côtiers. Aléas versus aménités, le cas particulier des territoires littoraux [Online]. Available at: https://journals.openedition.org/cybergeo/32249.

Dachary-Bernard, J., Gaschet, F., Lyser, S., Pouyanne, G., Virol, S. (2011). L'impact de la littoralisation sur les marchés fonciers. Une approche comparative des côtes Basque et Charentaise. *Économie et statistique*, 444/445, 127–154.

Décamps, A. and Gaschet, F. (2013). La contribution des effets de voisinage à la formation des prix du logement. Une évaluation sur l'agglomération bordelaise. *Revue économique*, 64(5), 883–910.

Dejean, A., Hild, A., Rotaru, R., Sasso, M., Vuillemier-Papaloizos, D., Clergeau, P. (2019). Des leviers d'action pour favoriser la biodiversité urbaine dans le cadre du processus de métropolisation [Online]. Available at: https://journals.openedition.org/cybergeo/32578.

Desrivierre, D. (2017). D'ici 2050, la population augmenterait dans toutes les régions de métropole. Report, INSEE, Paris.

Douvinet, J., Defossez, S., Anselle, A., Denolle, A.-S. (2011). Les maires face aux plans de prévention du risque inondation. *L'Espace géographique*, 40(1), 31–46.

Dubé, J., Des Rosiers, F., Thériault, M. (2011). Segmentation spatiale et choix de la forme fonctionnelle en modélisation hédonique. *Revue d'économie régionale urbaine*, 1, 9–37.

Freeman III, A.M. (1979). Hedonic prices, property values and measuring environmental benefits: A survey of the issues. *The Scandinavian Journal of Economics*, 154–173.

Halleck Vega, S. and Elhorst, J.P. (2015). The SLX model. *Journal of Regional Science*, 55(3), 339–363.

Hubert, G., Capblancq, J., Barroca, B. (2003). L'influence des inondations et des documents réglementaires sur le marché foncier en zone inondable. *Annales des ponts et chaussées*, 105, 32–39.

IPCC (2014). Climate change 2014: Impacts, adaptation, and vulnerability. Part A: Global and sectoral aspects. Report, Intergovernmental Panel on Climate Change, Geneva.

IPCC (2019). IPCC special report on the ocean and cryosphere in a changing climate. Report, Intergovernmental Panel on Climate Change, Geneva.

IPCC (2021). Climate change 2021: The physical science basis. Contribution of Working Group I to the Sixth Assessment Report of the Intergovernmental Panel on Climate Change [Online]. Available at: https://report.ipcc.ch/ar6/wg1/IPCC_AR6_WGI_FullReport.pdf.

Kousky, C. (2010). Learning from extreme events: Risk perceptions after the flood. *Land Economics*, 86(3), 395–422.

Kousky, C., Michel-Kerjan, E.O., Raschky, P.A. (2018). Does federal disaster assistance crowd out flood insurance? *Journal of Environmental Economics and Management*, 87, 150–164.

Lancaster, K.J. (1966). A new approach to consumer theory. *Journal of Political Economy*, 74, 132–157.

Longuépée, J. and Zuindeau, B. (2001). L'impact du coût des inondations sur les valeurs immobilières : une application de la méthode des prix hédoniques à la basse vallée de la Canche. *Cahiers du GRATICE*, 21, 143–166.

Mauroux, A. (2015a). Exposition aux risques naturels et marchés immobiliers. *Revue d'économie financière*, 117(1), 91–103.

Mauroux, A. (2015b). L'information préventive améliore-t-elle la perception des risques ? Impact de l'information acquéreur locataire sur le prix des logements. Working paper, FAERE – French Association of Environmental and Resource Economists.

Mills, E. (1967). An aggregate model of resource allocation in a metropolitan area. *American Economic Review*, 57, 197–211.

Muth, R.F. (1969). *Cities and Housing: The Spatial Pattern of Urban Residential Land Use*. University of Chicago Press, Chicago/London.

Næss, L.O., Bang, G., Eriksen, S., Vevatne, J. (2005). Institutional adaptation to climate change: Flood responses at the municipal level in Norway. *Global Environmental Change*, 15(2), 125–138.

November, V. (1994). Risques naturels et croissance urbaine : réflexion théorique sur la nature et le rôle du risque dans l'espace urbain. *Revue de géographie alpine*, 82(4), 113–123.

Peacock, W.G., Brody, S.D., Highfield, W. (2005). Hurricane risk perceptions among Florida's single family homeowners. *Landscape and Urban Planning*, 73(2), 120–135.

Poussin, J.K., Botzen, W.J.W., Aerts, J.C.J.H. (2013). Stimulating flood damage mitigation through insurance: An assessment of the French CatNat system. *Environmental Hazards*, 12(3/4), 258–277.

Rosen, S. (1974). Hedonic prices and implicit markets: Product differentiation in perfect competition. *Journal of Political Economy*, 82(1), 34–55.

Senil, N. (2021). Vers un tournant rural en France ? [Online]. Available at: https://theconversation.com/vers-un-tournant-rural-en-france-151490.

Siegrist, M. and Gutscher, H. (2008). Natural hazards and motivation for mitigation behavior: People cannot predict the affect evoked by a severe flood. *Risk Analysis*, 28(3), 771–778.

Surminski, S., Aerts, J.C.J.H., Botzen, W.J.W., Hudson, P., Mysiak, J., Pérez-Blanco, C.D. (2015). Reflections on the current debate on how to link flood insurance and disaster risk reduction in the European Union. *Natural Hazards*, 79(3), 1451–1479.

Tobin, G.A. and Montz, B.E. (1989). Catastrophic flooding and the response of the real estate market. *The Social Science Journal*, 25(2), 167–177.

Tobin, G.A. and Montz, B.E. (1990). Response of the real estate market to frequent flooding: The case of Des Plaines, Illinois. *Bulletin of the Illinois Geographical Society*, 33(2), 11.

Tobin, G.A. and Newton, T.G. (1986). A theoretical framework of flood induced changes in urban land values. *Journal of the American Water Resources Association*, 22(1), 67–71.

Travers, M., Bonnet, E., Chevé, M., Appéré, G. (2009). Risques industriels et zone naturelle estuarienne : une analyse hédoniste spatiale. *Économie & prévision*, 190/191(4), 135–158.

United Nations (2019). World urbanization prospects: The 2018 revision. Report, United Nations, New York.

4

The Ecological Restoration of Estuaries: Protection of People and Combating the Erosion of Biodiversity

Estuaries are critical refuges for plant and animal species that have commercial, recreational and cultural value. A wide variety of habitats are associated with estuaries, such as intertidal mudflats, sandy beaches, rocky shores, salt marshes and tidal channels. These different habitats are home to remarkable biodiversity adapted to these highly changing environments. The mixture of fresh and salt waters, the variations of water levels due to the tide, the temperatures and direction of the flow at each tide, make these original environments habitats of great ecological importance.

Estuaries have always been communication routes between the sea and the continent: 22 of the 32 largest cities in the world are located on estuaries (Hong Kong, Bangkok, Manila, London, New York, etc.). Human settlements and activities in estuaries have led to degradation of estuaries and aquatic resources, as well as the terrestrial biodiversity associated with these environments.

Each activity potentially affects the quality of the environment, but it is often the interaction between various human activities that produces the most significant impacts on the functioning, productivity and biodiversity of estuaries.

Chapter written by Mario LEPAGE, Michael ELLIOTT, Cécile CAPDERREY and Henrique CABRAL.

Recently, there has been growing ecological awareness in the population, as well as in public decisions, through commitments in favor of nature. These national and international actions to promote ecological restoration and biodiversity echo what experts are terming "the sixth extinction", a mass extinction of life on Earth greatly accelerated by human activities (Ceballos et al. 2015).

Issues of habitat and species protection and restoration are initially linked to human well-being which emanate from ecosystem services (ES). Estuaries, such as coastal areas, perform many important functions for living organisms and produce cumulative ES, goods and societal benefits due to their position as an interface between continental aquatic and terrestrial ecosystems and the sea (Barbier et al. 2011).

Ecosystem services (ES) and societal goods and benefits (SG&B) are often represented by a cascade of effects, in which biophysical structures and processes (habitats or primary production) influence key functions (regulation, production, information), the latter playing a role in providing services (erosion protection, fish production), which in turn provide goods and benefits to humans (fishing, welfare, health, safety), and ultimately have some market or non-market value. These SG&Bs may then have monetary or cultural value. The estimated value of the goods and benefits to humans will influence decision-making processes and the implementation of regulations and ecosystem management measures. This means that positive effects may cascade, but negative effects may also follow the same trajectory.

Among these ES provided by estuaries are provisioning services (food, freshwater, forest resources and energy), regulating and maintaining services (climate regulation, flood regulation, disease regulation and water purification), and cultural benefits (aesthetic, spiritual, educational and recreational). However, while the digging of navigation channels and the draining of wetlands have reinforced developments for human activities, these actions have also contributed to the reduction of aquatic habitats for species that depend on them. The reduction of intertidal wetlands has sometimes even led to the erosion of local biodiversity, and thus to a loss of ES.

Estuarine restoration takes place in the context of climate change, which has effects at all levels of the estuarine system. In order for restoration to be successful, it is necessary to understand the history and trajectory that the ecosystem has experienced over the centuries. This enables better identification and understanding of the objectives of restoration and the means to achieve them by taking into account the ecological functioning and the services provided to humans.

This chapter presents the causes behind the degradation of estuaries and proposes restoration options to limit the risks to biodiversity and the risks for coastal residents in the context of climate change and rising sea levels.

4.1. Habitats, biodiversity and ecosystem services

Ecosystem degradation and resulting biodiversity loss threaten the flow of ES and hence the resultant delivery of societal benefits. Ecosystem conservation and restoration assisted by eco-engineering, more recently termed nature-based solutions, address these threats. Eco-engineered restoration aims to work with nature to address the challenges posed by the pressures of human activities and climate change. Coordinated action is needed so that lost habitats can be successfully recreated. The European Union (EU) Biodiversity Strategy sets a target that by 2030, ecosystems and their services should be maintained and enhanced through the provision of green infrastructure.

An objective assessment of activities and pressures on estuaries is essential to evaluate the impact of degradation. This starts with a good understanding of biogeochemical and geomorphological processes, on the one hand, and biotic functions, on the other. This makes it possible to verify that the basic processes are functional, before undertaking work to restore species or habitats. However, in order not to delay restoration, actions to limit or even eliminate pressures on estuaries should be undertaken without delay. Atkins et al. (2011) proposed a policy framework to identify and focus on the options available to estuary managers to create and maintain ES and SG&Bs. In addition, Atkins et al. (2015) proposed a set of quantitative indicators for ES and SG&Bs for estuarine and coastal areas, with the assumption that once these indicators are defined, monitoring can be conducted to ensure that the goals have been met.

4.2. Causes of the ecological degradation of estuaries

Due to their position at the river–sea interface, estuaries are subject to multiple stressors related to upstream human activities, such as water abstraction that reduces flows, loss of connectivity due to the construction of dams and dikes, and pollutant flows from various sources. The main anthropogenic impacts are caused by urbanization and industrialization, the artificialization of riverbanks, the heating of water by the cooling circuits of power plants, dredging activities associated with maritime traffic and recreational facilities, and the pressures of agricultural activities. In estuaries, water circulates in both directions due to the tide. Therefore, pressures can also arise from the sea during accidental pollution, shipwrecks, or urban or industrial discharges from cities or outlying areas. Most human pressures are in addition to the natural stressors that shape estuarine biological communities.

4.2.1. *Effects of rising sea levels*

As sea levels rise, the increased frequency and duration of flooding threatens life, infrastructure and many ESs. It tends to restrict intertidal areas as many estuaries are protected by dikes and levees, causing the coastline to narrow (termed "coastal squeeze"). Rising water levels also increase coastal erosion and the destruction of shallow, quiet habitats for juvenile fish life stages, as well as adjacent coastal habitats. A decrease in coastal vegetation is expected with these new water level conditions, thus affecting habitats, as well as the shore protection function and the damping effect to waves on the coast.

4.2.1.1. *Intertidal habitats*

For the ichthyofauna, the intertidal mudflats act as a nursery for the principal fish species, which the larvae of these species reach to carry out part of their life cycle. The expected effect of rising sea levels in estuaries is compression of intertidal areas and habitat loss in areas of high productivity such as mudflats. The development and maintenance of intertidal habitats (mudflats, salt marshes) are dependent on hydrological factors that condition the hydroperiod (frequency, duration, intensity and water height of immersions) and sediment deposition, as well as the spread of propagules on the foreshore. The quantity and frequency of the inflow play a key role in the development of vegetation. Too much sediment accumulation, for example, can bury seeds and smother seedlings, while too much erosion, in addition to destabilizing sediments, can influence the rate of seed release from sediments (Spencer and Harvey 2012). The relationships between morphology, hydroperiod, sediment composition, as well as microtopography condition the establishment and maintenance of plant species, and thus primary production and habitat development for fish and benthic invertebrates. Rising sea levels, in conjunction with climate change, will have effects on each of the above parameters, and thus on intertidal habitats overall.

4.2.1.2. Subtidal habitats

Seaweed and seagrass beds contribute to the fight against coastline erosion. By influencing local hydrodynamics, these macrophyte beds attenuate current speeds and wave energy. Susceptible to being impacted by the decrease in light inherent in turbidity, seagrass beds can experience a change in their lower boundary, and their loss can lead to erosion phenomena felt on a large scale (Walter et al. 2020). Vegetated subtidal habitats increase bottom roughness and help limit tidal and salinity front penetration into the estuary. As with intertidal vegetation, subtidal vegetation acts on energy dissipation for greater coastal protection from erosion.

4.2.2. Effects of anthropogenic pressures on biodiversity

Estuaries are not particularly species-rich, mainly due to their extremely high natural variability but they do support greater populations than other ecosystems (Elliott and Whitfield 2011). They are among the ecological systems most affected by human activities (Lotze et al. 2006). Although their species are generally tolerant of a wide range of natural stressors and human pressures (Elliott and Quintino 2007), a large number of studies have documented the effects of anthropogenic pressures on estuarine biodiversity and ES.

Habitat loss through land appropriation, the filling and draining of estuarine marshes, is a notable pressure worldwide and has a significant impact on species diversity and estuarine functioning. Most estuarine species exhibit spatial segregation being more abundant in some habitats than others. As habitats are not equally affected by these pressures, effects on estuarine community structures and ecosystem functioning also result from these anthropogenic impacts. Among the most impacted habitats are intertidal areas, salt marshes, mangroves, seagrasses and biogenic reefs (Mumby et al. 2014). These habitats, generally rich in benthic invertebrates, are the primary food resources for a wide variety of species (including fish and birds). In addition, changes can affect connectivity between habitats, the nursery function for fish or overwintering areas for birds, and even the primary productivity of these systems. A study in northern Portugal (the Lima estuary) estimated that 70% of its intertidal area and 88% of its salt marshes have been lost since the 1930s, and that the most attractive habitats for fish have declined by 48%. This loss of salt marsh and intertidal areas cannot be compensated for by the gain in deep subtidal habitats, given their different functional roles for fish and invertebrates (Amorim et al. 2017). In the Seine estuary, estimates of loss of suitable habitat for juvenile fish represent a 42% decrease in nursery capacity for sole, which is one of the primary fishery resources in this geographic area (Rochette et al. 2010).

Dam construction, dredging, sediment and water extraction have altered river flows and the hydrodynamics of estuarine systems. One of the best-documented impacts of river regulation and the construction of physical barriers in rivers is the effect on diadromous migratory fish. These fish species depend on environmental cues to initiate migratory movements, and their ability to reach spawning grounds that are generally located upstream and therefore inaccessible due to dams. Changes in river flow affect temperature and saline estuarine gradients, which determine the spatiotemporal distribution of all estuarine organisms (Barletta et al. 2005). For the Gironde estuary (France), changes in the distribution and abundance of species, namely copepods and fish, have been highlighted and linked to increasing salinity of estuarine waters (Pasquaud et al. 2012). There are also other impacts resulting from water abstraction, such as for power plant cooling, industry and agriculture, which can have significant impacts, such as mortality of individuals trapped in water pumping devices.

Urban, industrial and agricultural development has introduced a wide variety of pollutants into estuaries. Pollution can result in mass mortalities or the decline or loss of the most sensitive species. Nutrient-rich wastewater from urban areas and farm run-off has been responsible for oxygen depletion in several estuaries around the world, with negative consequences for biological communities. In the Tyne estuary (on the east coast of England), the proportion of small pelagic fish has increased relative to benthic fish, due to the presence of near-bottom deoxygenation zones (Hall and Frid 1997).

In the Tagus estuary in Portugal, pollution has had a negative effect on the fish nursery function, with a strong decline reported for some fish species (Costa and Cabral 1999). Heavy metals or persistent organic pollutants tend to be accumulated by organisms (bioaccumulation), which is particularly dangerous if the species is used for human consumption. In the Scheldt estuary (Netherlands–Belgium), the analysis of pea and invertebrate tissues showed high levels of contaminants, such as heavy metals, polychlorinated biphenyls (PCBs), polybrominated diphenyl ethers (PBDEs) and organochlorine pesticides (OCPs), and presented potential consequences for human health beyond the effects on the contaminated species. The effects of other types of pollution, such as underwater noise or anthropogenic light, have been examined more recently, with these primarily impacting the behavior and stress levels of organisms (Roberts et al. 2016).

The introduction of non-indigenous species (NIS) into estuaries has long occurred but has intensified in recent decades. Harbors and shipping traffic, as well as aquaculture, are among the activities responsible for the introduction of NIS. In Europe, an example of an introduced fish that has disrupted the ecological balance of estuaries is the catfish (*Silurus glanis*) (Savini et al. 2010). This is a major predatory fish, competing with the native species and changing habitat structure. Another example is the goldfish (*Carassius auratus*), an ornamental species, considered the most widespread NIS fish in Europe. Although originally a freshwater species, its distribution has spread from the oligohaline zone to other estuarine areas and it has changed the abundance and composition of estuarine fish assemblages (Tweedley et al. 2017). Certain crustacean and bivalve NIS have also had impacts on estuarine communities and ecosystem functioning. For example, the zebra mussel (*Dreissena polymorpha*), an invasive bivalve in the Hudson estuary (USA), has caused changes in populations of pelagic species such as American shad (*Alosa sapidissima*) through alteration of planktonic resources (Strayer et al. 2004). Green crab (*Carcinus maenas*), often the most abundant crab species in European estuaries, is considered an NIS in other geographic areas (e.g. North America or Australia), where it has caused changes in seagrass habitats and fish assemblages. In Europe, since its introduction over a century ago, the Chinese crab (*Eriocheir sinensis*) has colonized several estuaries and caused negative effects on local communities.

Climate change has also had several impacts on estuarine assemblages. Multiple physical and chemical changes have been recorded in estuaries, including temperature, pH, salinity and dissolved oxygen concentration. These changes affect the environmental gradients that characterize estuarine systems and determine the spatial and temporal distributions of different fish and shellfish species. In Australia, environmental changes affecting biodiversity have been tracked for many estuarine systems (more than 150) and have been reported at a very impressive rate: estuarine summer water temperature has increased by an average of 0.2°C per year, and pH has decreased by about 0.5 pH units in 12 years (Scanes et al. 2020). Documented biodiversity changes include colonization by tropical marine species, reduced species richness and increased risk of algal blooms, among others (Scanes et al. 2020). For several European estuaries, climate-induced changes have also been reported, notably for the Gironde, where changes are observed in planktonic and fishery communities (Chevillot et al. 2019). This is also the case for the Mondego estuary in Portugal, where the analysis of a two-decade planktonic community dataset highlighted the decrease of freshwater species, and the increase of small cosmopolitan copepods and gelatinous zooplankton (Falcão et al. 2012).

4.3. Ecological restoration of estuaries for the protection of biodiversity

Restoration of degraded aquatic ecosystems requires a coordinated and holistic approach involving ecohydrology and ecological engineering (eco-engineering) (Wolanski and Elliott 2016). This approach emphasizes a systems approach (Figure 4.1) that encompasses the continuum of ecosystem structure and biodiversity, which are underpinned by the marine and coastal physicochemical system, which then gives rise to ecological functioning. We will refer to this whole as "ecology" for ease of reference in the text that follows. If successful, the marine and coastal system produces ESs that, after the input of complementary human assets and capital, provide SG&Bs. For example, if tides and sediments are maintained, the ecosystem can maintain its flora and fauna. However, the ecology can be degraded by human activities and pressures, reducing ESs and SG&Bs. Restoring the ecology through nature-based solutions and eco-engineering will, in turn, enhance ES and SG&B.

Passive restoration is defined as the ability to remove the activities and their pressures (defined as the mechanisms of change) that cause degradation, thereby allowing the ecology to recover on its own. In contrast, active restoration involves human interventions through nature-based solutions and ecological engineering, also called eco-engineering.

Eco-engineering has been divided into two types: type A, which involves restoring the habitat to allow the ecology and its species to thrive; and type B, which involves human-mediated enhancement of ecological functioning and species by restoring flora and/or replenishing fauna (Elliott et al. 2016). A type A approach focuses first on changing the physical form of the areas to be restored, that is, the physiography and geomorphology, and perhaps manipulating water flows (fresh or salt), to create ecological niches and habitats. In a type B approach, ecological engineering consists of restocking, replanting, which also creates habitats, and enhancing the interactions between biota and their environment. In turn, these interactions will increase the resistance and resilience of an ecosystem to respectively adapt and recover from the effects of human pressures. Eco-engineering is considered here as a holistic approach that uses scientific knowledge to modify and achieve ecological goals in certain parts or in the whole coastal and marine system. This approach is the essence of nature-based solutions. However, ecosystem management through eco-engineering and nature-based solutions must be complemented with the regulation of human activities on land and in the aquatic environment.

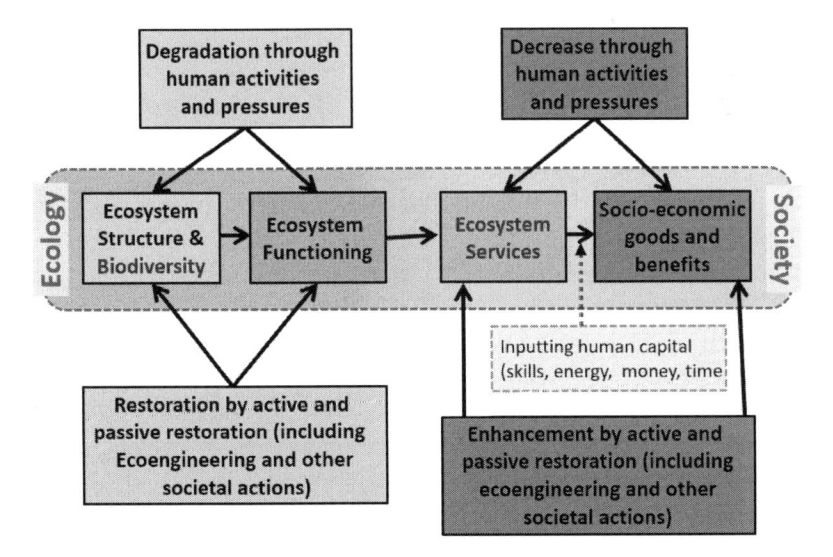

Figure 4.1. *The systemic basis of ecological restoration. For a color version of this figure, see www.iste.co.uk/salles/estuarinecities.zip*

Coastal areas around the world, and particularly in Europe, are subject to increasing urbanization and industrialization, the increased use of resources such as water, energy, space and food, and the decreased resistance and resilience to stresses such as climate change (Defeo and Elliott 2021). It is unrealistic to expect that eco-engineering and nature-based solutions can fully recreate or restore estuarine ecosystems, especially those with large human populations, such as Europe. Nevertheless, the goal is to create ecosystems with the attributes of the original systems and to aim for improved ecological functioning. These restoration goals provide benefits to the economy and to societal security, in addition to achieving some ecological balance by addressing some historical changes, such as wetland removal (a so-called *win–win–win* strategy). It provides low-cost technologies to mitigate the impact of human activities on transitional waters, coasts and marine areas, to use and enhance the natural capacity of water bodies to absorb or treat excess nutrients and pollutants, and to increase resilience to global stressors, such as climate change (Wolanski and Elliott 2016).

The successful use of ecological engineering and nature-based solutions for ecosystem and habitat remediation and restoration is based on our understanding of the natural functioning within a socioeconomic framework. In essence, it is a holistic approach governed by and for the various estuarine stakeholders, using the best available techniques and technologies, and within a cost-effective and beneficial framework. These approaches recognize that successful ecosystem restoration is generally a balance between natural regeneration and recovery using appropriate management measures, including nature-based solutions, such as eco-engineering techniques, combined with stressor removal. The effectiveness of natural and social system restoration is measured in terms of ES and SG&B, especially those resulting from the creation of sustainable ecological structure and functioning (Turner and Schaafsma 2015). Similarly, it appears necessary to assess SG&B at the habitat scale before moving to a broader regional scale. It is therefore necessary to know how society and stakeholders in the wider estuary understand and perceive habitats, their ecological structure and functioning, and the ES and SG&B they produce. However, it is first necessary to determine the loss of these ES and SG&B resulting from environmental degradation, due to single and/or multiple pressures.

To date, restoration practices have been implemented primarily at the local level and in isolation, at a site, rather than being integrated into the management of the ecosystem as a whole. This hinders recognition of the additional benefits of establishing networks of restoration measures at regional and European levels.

Ecosystem restoration provides greater resilience and protection from flooding, and can reduce protection costs while creating more habitat and protecting species. Lessons learned from existing restoration actions help identify clear criteria for selecting the most cost-effective sites and techniques for future programs, to maximize societal benefits while reducing costs.

4.3.1. *Active and passive restoration*

The model in Figure 4.2 emphasizes that restoration can be passive, removing human-induced stressors and allowing the natural system to recover; or it can be active, in which habitats are created where fauna and flora are enhanced through restocking or replanting. After degradation, habitats can be restored which improves the degraded state, but restoration does not necessarily return habitats to their original state, whereas newly created ecosystems, such as levee setback sites, can provide valuable ES. Restoration may involve returning to the original state or moving away from the ecosystem quality recovery trajectory, depending on the societal benefits required as a result of ES. Some limitations to restoration are unavoidable, given the potential irreversible effects of degradation, upstream changes in the catchment or environmental constraints (e.g. projected future changes due to climate change), as well as socio-economic constraints, such as available resources and the extent of community and stakeholder support for the estuary.

Active restoration activities in a habitat or ecosystem can be considered to have both a direct and indirect footprint, or area of influence. For example, removing a barrier to address disturbances in the hydrography of an area has a direct effect on the area where the barrier was located, but it also affects the hydrography downstream and the ecology upstream and downstream, including the large-scale movements of migratory fish through estuaries. Similarly, the creation of a feeding, breeding and resting wetland for birds, or a nursery area for fish, can have an indirect effect on their populations far away.

An applied pragmatic approach is required for planning and monitoring ecosystem restoration. Practitioners, including national and regional planners, are constantly confronted with the causes and consequences of environmental change, and must therefore determine management responses to address restoration challenges. This cause–consequence–response chain must also take into account different spatial and temporal scales. Consideration of the full sequence is key to problem solving in order to holistically assess and address causes, consequences and responses to change. For example, the restoration of transitional waters, initially subject to increased nutrient runoff from agricultural lands, would avoid symptoms of coastal eutrophication that impact local tourism and aquaculture activities. Similarly, removing or limiting the adverse effects of industrial discharges in an estuary removes anoxic water quality barriers for migratory fish species that spawn in freshwater. This has a cost–benefit to both the recreational and commercial fishing sectors, beyond a general improvement in water quality. As such, determining the management effectiveness of restoration measures requires objectives to be evaluated together with quantitative indicators termed SMART (Specific, Measurable, Achievable, Realistic and Time-bound) (Buijse 2015).

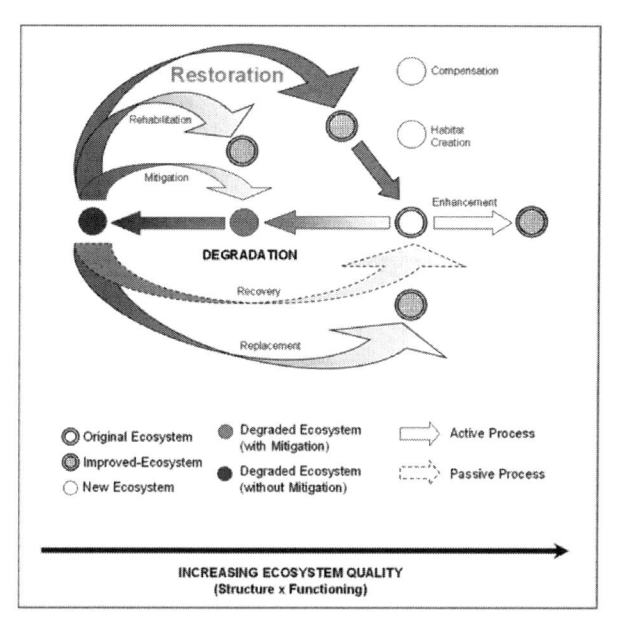

Figure 4.2. *Active and passive responses to ecosystem degradation (source: Elliott et al. (2007)). For a color version of this figure, see www.iste.co.uk/salles/estuarinecities.zip*

4.4. Examples of ecological restoration in estuaries

There are many examples of estuarine restoration around the world showing the diversity of possible actions to achieve different objectives (Boerema and Meire 2017). We have chosen to present here a selection of restoration case studies that take into account ecological, economic and cultural aspects, while focusing on the protection of people and property.

4.4.1. *The marshes of Mortagne-sur-Gironde (France)*

Figure 4.3. *Aerial view of the Mortagne marsh in 1957, before the construction of the polder (source: IGN Géoportail, France)*

In December 1999, cyclone Martin (with winds over 200 km/h) damaged the dike of the Mortagne-sur-Gironde polder in Charente-Maritime. This dike had been created in 1966 to develop cereal cultivation on a salt marsh in the Gironde estuary (see Figure 4.3 for the view of the site before reclamation). Very soon after the storm, the decision was made to reinforce the back dike to protect homes and other agricultural land, and not to repair the "breach" in the dike. The subsequent flooding of the previously created polder began by creating a regressive erosion channel within the polder, and a thin layer of fresh sediment covered the still visible traces of the corn rows cut the previous autumn (Figure 4.4). From there, a succession of passive restoration of the marsh began. After months of regular flooding of the polder through the single breach (gap), vegetation of halophytic and sub-halophytic plants (Decreton 2009; Allou 2016), typical of salt marshes, gradually erased the traces of corn. The configuration of the site – a path divided the polder into two parts – led to the upstream part of the path remaining under water, while the downstream part emptied almost entirely at each tide. During this period, a large number of benthic invertebrates (polychaete worms, crustaceans, gastropods and bivalves) settled in both the underwater and intertidal sections. Wild oysters came to colonize the part in permanent water. Gradually, erosion removed the path, which was acting as a threshold, and the marsh completely emptied at each tide. Today, 20 years after the first flooding of the polder, the marsh is vegetated with tallgrass vegetation, and there is little water left at low tide when compared to these first years (Figure 4.4). The vegetated portions are inundated during large equinox tides and remain wetland most of the time. Although this is a passive return to nature of a polder that had been claimed from the sea, the habitats lost during the original land-claim (reclamation) were only partially recovered with this new ecosystem. Nesting functions for several passerine species are present, but feeding functions for shorebirds and fish are reduced to the part that remains in water. Although no measurements have been made to estimate the self-purification of the Mortagne marsh, experiments on other intertidal marshes (Simas and Ferreira 2007) have shown that this type of habitat reduces the nutrient load entering the marsh.

Figure 4.4. *Aerial views of the Mortagne marsh. Top left, the polder before the 1999 cyclone; top right, view at low tide in 2002; bottom left, the marsh in 2006 at high tide; bottom right, the marsh in 2018 at mid-tide (source: IGN Géoportail, France). For a color version of this figure, see www.iste.co.uk/salles/estuarinecities.zip*

4.4.2. *Mondego estuary (Portugal)*

The Mondego estuary is a small warm-temperate mesotidal system located on the west coast of Portugal (4008 N; 850 W). With an area of 8.6 km² and a length of 21 km, it crosses the lower Mondego valley, surrounded by 150 km² of valuable agricultural land. Inside the estuary, at the mouth of the river, the stream divides into two arms encircling the island of Morraceira. The northern arm reaches depths of 5–10 m at high tide, while the southern arm is limited to a depth of 2–4 m. This southern arm is composed of approx. 75% intertidal mudflats, whereas in the northern arm, less than 10% of this type of habitat is found. The tidal range is different for the two arms, being 2–3 m in the north and 1–3 m in the south. This difference is primarily due to the largely flooded upstream areas of the southern arm, which result in differences in water flow. The estuary has high economic value since it is home to many industries, including salt flats, aquaculture facilities, commercial and fishing ports, and agricultural areas, which cause many physical impacts and nutrient discharges into the area. The northern arm receives most of the marine traffic and is frequently dredged.

The Mondego estuary has been subject to strong anthropogenic pressures over the years and can be considered a model of historical human exploitation. These pressures have resulted in hydro-morphological changes and, consequently, ecological decline. In an attempt to reverse this situation, a strategy including several mitigation measures was first designed in 1997–1998, culminating in more intense action in 2006. Over several decades, multiple anthropogenic stressors (e.g. eutrophication, dredging and physical alterations) have resulted in several changes in the Mondego estuary. The most serious phenomenon was eutrophication of the southern arm in the early 1990s. This led to a decline in water quality, due to the closed communication between the two arms in the upper section caused by several physical modifications within the estuary (Castro et al. 2016). The obstruction led to an increase in water residence time and nutrient concentration, which favored the proliferation of macroalgae (*Ulva* sp.) and the regression of the macrophyte (seagrass) cover (*Z. noltei*) in this part of the estuary, from 15 ha in 1986 to 0.02 ha in 1997.

Some mitigation measures were implemented in the late 1990s to reverse this situation. The goals were to improve water quality, and reduce nutrient loading and water residence time in the southern arm. Two main measures were implemented:

– reducing freshwater and nutrient input from the Pranto River into the southern arm by diverting it upstream to the northern arm where the dilution factor was greater;

– improving water flow by partially restoring (~1 m wide) the upstream connection between the southern and northern arms for limited periods (1 h 30 min–2 h before and after peak high water).

After this intervention, the system showed signs of improvement: the seagrass beds partially recovered and macro-algae blooms were no longer detected. Due to these improvements after the implementation of the first measures, the Portuguese Environmental Agency adopted a second, larger-scale intervention plan for the system. Therefore, in 2006, the restoration of the communication between the two arms was made permanent in order to increase the water flow and reduce the residence time of the water in the southern arm. The Pranto River detour was achieved through engineering; however, the restoration of water quality and seagrass beds is the result of the effect of nature-based solutions (Dolbeth et al. 2011).

4.4.3. *Scheldt estuary (Belgium)*

Following the catastrophic floods of the 1950s that resulted in many victims along the Scheldt, and following many other flood events in the estuarine area, the governments of Belgium and the Netherlands have been working since the early 1990s on a flood control plan, called Sigmaplan. The primary objective was flood control, but over the years ecological considerations have become increasingly important.

Wetland projects have been developed all along the estuary, from the mouth of the Scheldt to upstream of Antwerp. Several options were chosen to create flood control areas (FCA), managed realignment (MR) sites and temporary storage areas for overflow water, called controlled reduced tide (CRT) areas. All of these techniques for hydro-physical modification of the estuary actually converge on an initial situation where numerous overflow areas existed, before the river was dammed and its banks were occupied by port infrastructures and urban and industrial areas. These developments give the river more room to flow, while ensuring that the water overflows preferentially in certain areas to protect homes and people. The latest project is the depoldering of the Hedwige/Propser polder on the Dutch–Belgian border, which moves the dike back to create 600 ha of tidal wetland, which is returned to the Scheldt (Figure 4.5). The impoundment of this new area with a very large storage capacity is part of the Sigmaplan. Work on this site is scheduled to be completed in 2023.

Smaller works, useful for increasing the functioning for aquatic communities and birds, were carried out by the retreat of the Lillo dike approximately 20 km downstream of Antwerp. This 10-ha project has proven to be very valuable as a breeding site for many bird species. The site is developed as a promenade for visitors who enjoy estuarine landscapes and birds, with access to certain areas being restricted during the nesting period to avoid disturbance of the nests. The intertidal mudflats thus created are productive zones for benthic invertebrates and for the microphytobenthos, which develop there and contribute to the local food web.

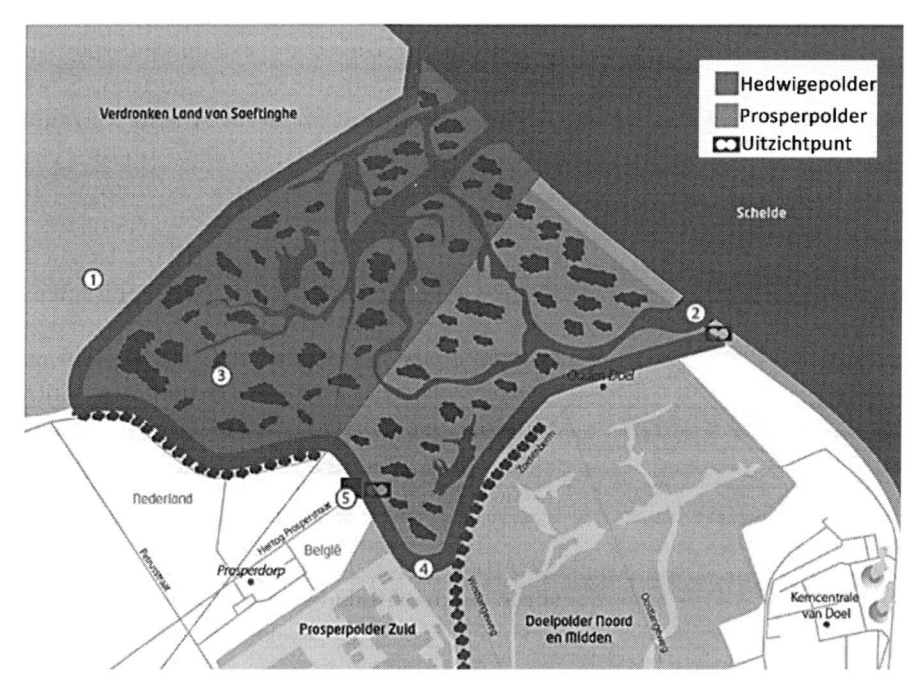

Figure 4.5. *Dike retreat on the Scheldt estuary, the Hedwig-Prosper polder (from the INTEREG Polder2C project). For a color version of this figure, see www.iste.co.uk/salles/estuarinecities.zip*

The construction of CRT areas combines the possibility of large-volume water storage, recreation of natural habitat and agricultural use, while at the same time providing space for recreational activities for nature. The first example of this type was the pilot project in Lippenbroek (10 ha), built on a diked agricultural area. This system has proven to be useful on several levels and is one of the *win–win–win* systems, combining flood risk mitigation (and hence public safety), economic development and the creation of a viable ecosystem. Indeed, the CRTs fulfill their flood control role well by temporarily storing a large volume of water that can prevent uncontrolled overflows elsewhere. The tourist attraction of the site creates a regional asset as a birdwatching site, a place for environmental education for schoolchildren, and a place for walks or sporting activities such as cycling or running. The maintenance of a tide inside the CRT allows the development of adapted flora that allows for a self-purifying action of the water that stays in the CRT. The area also serves as a habitat for many species of invertebrates, fish and birds, as well as small mammals that live around the wetland.

Figure 4.6. *Reversal of the dike at the Lillo site on the Scheldt estuary in Belgium. For a color version of this figure, see www.iste.co.uk/salles/estuarinecities.zip*

The Kruibeke polder, located immediately upstream of Antwerp and covering an area of 180 ha, is a good example of this CRT operation. Forestry activity is maintained on the highest parts, and grassy meadows extend into the water areas. The entire dike belt around the polder is developed for walking and picnic areas. A river branch runs from the river to the outlet of a small tributary of the Scheldt. All of these functions are provided at the same time as the objective of flood control and protection of people and property in case of an extreme event.

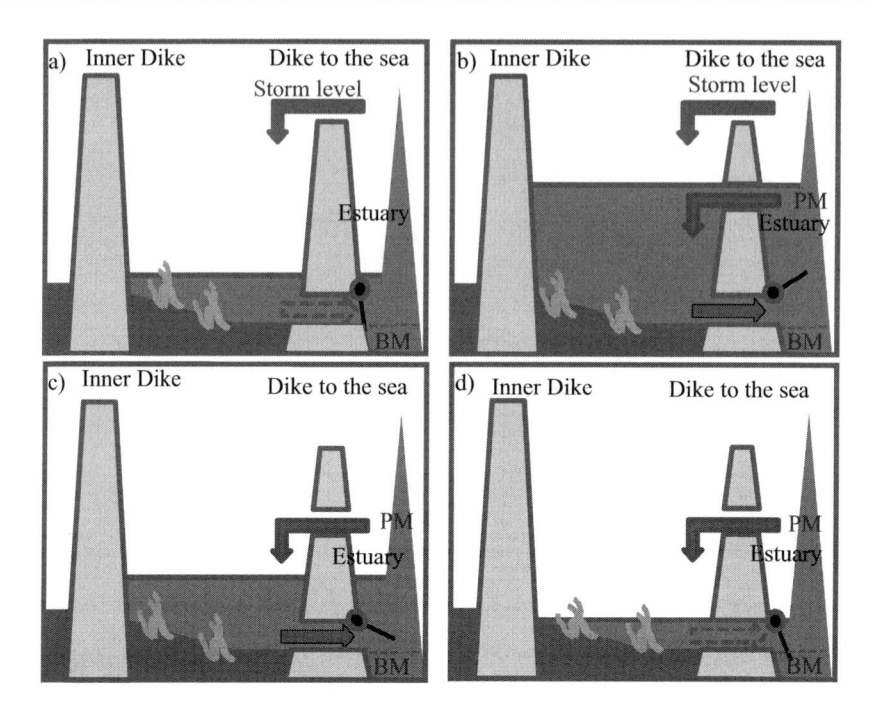

Figure 4.7. *Controlled reduced tide (CRT) operation. a) Storage area operation during floods. b) CRT operation during storms. c) Normal operation during spring tide. d) Normal operation during neap tide (modified from Esteves (2014)). For a color version of this figure, see www.iste.co.uk/salles/estuarinecities.zip*

4.5. Conclusion

The era of the artificialization and progressive destruction of estuaries is no longer acceptable to many citizens, as shown by the contestation of new industrial development projects on estuaries. In most cases, the point of no return has not been reached and there is hope of returning to a more natural state. The challenge for the ecological restoration of estuaries now consists of determining how, taken together, optimal combinations of restoration programs can enhance biodiversity and maximize the benefits for the safety of property and people. For example, resilience to rising sea levels or temperature increases due to climate change must be a priority. In addition, actions must be taken to better withstand changes in rainfall patterns and potential droughts. It is necessary to adapt to climate change, while also adapting the uses and the economy to these changes (triple benefit).

There is an urgent need to examine the effect of aquatic restoration projects on the major large-scale problems facing society, including water quality, nutritional status, primary production, hydro-morphological processes and biodiversity loss. Similarly, there is a need to address the impact of restoration at the regional and European scales on:

– the imbalance between the supply of and demand for ES and SG&B;

– how restoration successfully addresses environmental degradation in estuarine functioning;

– the upstream–downstream *continuum*.

From an ecological, societal and political perspective, estuarine areas are complex systems, given their multiple uses operating at different scales. It is therefore necessary to create and apply methods and indicators to evaluate the success of restoration efforts with respect to ecology, economics and society.

4.6. References

Allou, S. (2016). Renaturation d'un marais estuarien : réponses des poissons et des macrocrustacés à l'échelle des communautés et des individus. Suivi de la dépolderisation du marais de Mortagne. Mémoire de Master 2, Agrocampus Ouest, Rennes.

Amorim, E., Bordalo, A., Ramos, S., Elliott, M., Franco, A. (2017). Habitat loss and gain: Influence on habitat attractiveness for estuarine fish communities. *Estuar. Coast. Shelf Sci.*, 197.

Atkins, J.P., Burdon, D., Elliott, M., Gregory, A.J. (2011). Management of the marine environment: Integrating ecosystem services and societal benefits with the DPSIR framework in a systems approach. *Mar. Pollut. Bull.*, 62(2), 215–226.

Atkins, J.P., Burdon, D., Elliott, M. (2015). Chapter 5: Identification of a practicable set of indicators for coastal and marine ecosystem services. In *Coastal Zones Ecosystem Services: From Science to Values and Decision-Making Studies*, Turner, R.K. and Schaafsma, M. (eds). Springer, New York.

Barbier, E.B., Sally, D.H., Chris, K., Evamaria, W.K., Adrian, C.S., Brian, R.S. (2011). The value of estuarine and coastal ecosystem services. *Ecol. Monogr.*, 81, 169–193.

Barletta, M., Barletta-Bergan, A., Saint-Paul, U., Hubold, G. (2005). The role of salinity in structuring the fish assemblages in a tropical estuary. *Journal of Fish Biology*, 66(1), 45–72.

Boerema, A. and Meire, P. (2017). Management for estuarine ecosystem services: A review. *Ecol. Eng.*, 98, 172–182.

Buijse, T. (2015). REFORM: REstoring rivers FOR effective catchment Management REFORM EU FP7 [Online]. Available at: www.reformrivers.eu [Accessed 20 January 2023].

Castro, N., Félix, P., Neto, J.M., Cabral, H., Marques, J.C., Costa, M.J., Costa, J.L. (2016). Fish communities' response to implementation of restoring measures in a highly artificialized estuary. *Estuar. Coast. Shelf Sci.*, 67, 743–752.

Ceballos, G., Ehrlich, P.R., Barnosky, A.D., García, A., Pringle, R.M., Palmer, T.M. (2015). Accelerated modern human–induced species losses: Entering the sixth mass extinction [Online]. Available at: https://doi.org/10.1126/sciadv.1400253.

Chevillot, X., Tecchio, S., Chaalali, A., Lassalle, G., Selleslagh, J., Castelnaud, G., David, V., Bachelet, G., Niquil, N., Sautour, B. et al. (2019). Global changes jeopardize the trophic carrying capacity and functioning of estuarine ecosystems. *Ecosystems*, 22, 473–495.

Costa, M.J. and Cabral, H.N. (1999). Changes in the Tagus nursery function for commercial fish species: Some perspectives for management. *Aquatic Ecology*, 33(3), 287–292.

Decreton, T. (2009). Etude des fonctionnalités et des gains ichtyologiques liés à la dépoldérisation des zones humides estuariennes : le cas du marais de Mortagne-sur-Gironde. Master sciences de la terre, de l'eau et de l'environnement, Université François Rabelais, Tours.

Defeo, O. and Elliott, M. (2021). The "triple whammy" of coasts under threat – Why we should be worried! *Marine Pollution Bulletin*, 163, 111832.

Dolbeth, M., Cardoso, P.G., Grilo, T.F., Bordalo, M.D., Raffaelli, D., Pardal, M.A. (2011). Long-term changes in the production by estuarine macrobenthos affected by multiple stressors. *Estuar. Coast. Shelf Sci.*, 92, 10–18.

Elliott, M. and Quintino, V. (2007). The estuarine quality paradox, environmental homeostasis and the difficulty of detecting anthropogenic stress in naturally stressed areas. *Mar. Pollut. Bull.*, 54, 640–645.

Elliott, M. and Whitfield, A. (2011). Challenging paradigms in estuarine ecology and management. *Estuar. Coast. Shelf Sci.*, 94, 306–314.

Elliott, M., Burdon, D., Hemingway, K.L., Apitz, S.E. (2007). Estuarine, coastal and marine ecosystem restoration: Confusing management and science – A revision of concepts. *Estuar. Coast. Shelf Sci.*, 74(3), 349–366.

Elliott, M., Mander, L., Mazik, K., Simenstad, C., Valesini, F., Whitfield, A., Wolanski, E. (2016). Ecoengineering with ecohydrology: Successes and failures in estuarine restoration. *Estuar. Coast. Shelf Sci.*, 176, 12–35.

Esteves, L.S. (2014). *Managed Realignment: Is it a Viable Long-term Coastal Management Strategy?* Springer, Dordrecht.

Falcão, J., Marques, S.C., Pardal, M.A., Marques, J.C., Primo, A.L., Azeiteiro, U.M. (2012). Mesozooplankton structural responses in a shallow temperate estuary following restoration measures. *Estuar. Coast. Shelf Sci.*, 112, 23–30.

Hall, J.A. and Frid, C.L.J. (1997). Estuarine sediment remediation: Effects on benthic biodiversity. *Estuar. Coast. Shelf Sci.,* 44, 55–61.

Lotze, H.K., Lenihan, H.S., Bourque, B.J., Bradbury, R.H., Cooke, R.G., Kay, M.C., Kidwell, S.M., Kirby, M.X., Peterson, C.H., Jackson, J.B.C. (2006). Depletion, degradation, and recovery potential of estuaries and coastal seas. *Science*, 312, 1806–1809.

Mumby, P.J., Chollett, I., Bozec Y.-M., Wolff N.H. (2014). Ecological resilience, robustness and vulnerability: How do these concepts benefit ecosystem management? *Current Opinion in Environmental Sustainability*, 7, 22–27.

Pasquaud, S., Béguer, M., Larsen, M.H., Chaalali, A., Cabral, H., Lobry, J. (2012). Increase of marine juvenile fish abundances in the middle Gironde estuary related to warmer and more saline waters, due to global changes. *Estuar. Coast. Shelf Sci.,* 104, 46–53.

Roberts, L., Pérez-Domínguez, R., Elliott, M. (2016). Use of baited remote underwater video (BRUV) and motion analysis for studying the impacts of underwater noise upon free ranging fish and implications for marine energy management. *Marine Pollution Bulletin*, 112(1), 75–85.

Rochette, S., Rivot, E., Morin, J., Mackinson, S., Riou, P., Le Pape, O. (2010). Effect of nursery habitat degradation on flatfish population: Application to Solea solea in the Eastern Channel (Western Europe). *Journal of Sea Research*, 64(1–2), 34–44.

Savini D., Occhipinti–Ambrogi A., Marchini A., Tricarico E., Gherardi F., Olenin S., Gollasch S. (2010). The top 27 animal alien species introduced into Europe for aquaculture and related activities. *Journal of Applied Ichthyology*, 26, 1–7.

Scanes E., Scanes P.R., Ross P.M. (2020). Climate change rapidly warms and acidifies Australian estuaries. *Nature Communications*, 11, 1803.

Simas, T.C. and Ferreira, J.G. (2007). Nutrient enrichment and the role of salt marshes in the Tagus estuary (Portugal). *Estuar. Coast. Shelf Sci.*, 75(393).

Spencer, K.L. and Harvey, G.L. (2012). Understanding system disturbance and ecosystem services in restored saltmarshes: Integrating physical and biogeochemical processes. *Estuar. Coast. Shelf Sci.*, 106, 23–32.

Strayer, D.L., Hattala, K.A., Kahnle, A.W. (2004). Effects of an invasive bivalve (Dreissena polymorpha) on fish in the Hudson River estuary. *Can. J. Fish. Aquat. Sci.*, 61, 924–941.

Turner, R.K. and Schaafsma, M. (ed.) (2015). *Coastal Zones Ecosystem Services: From Science to Values and Decision Making*. Springer, Switzerland.

Tweedley J.R., Hallett C.S., Beatty S.J. (2017). Baseline survey of the fish fauna of a highly eutrophic estuary and evidence for its colonisation by Goldfish (*Carassius auratus*). *International Aquatic Research*, 9, 259–270.

Walter, R.K., O'Leary, J.K., Vitousek, S., Taherkhani, M., Geraghty, C., Kitajima, A. (2020). Large-scale erosion driven by intertidal eelgrass loss in an estuarine environment. *Estuar. Coast. Shelf Sci.*, 243, 106910.

Wolanski, E. and Elliott, M. (eds) (2016). *Estuarine Ecohydrology*. Elsevier, Amsterdam.

5

Sensemaking in the Face of Estuarine Flood Risk Mitigation

In *Metapolis, or the Future of Cities*, Ascher (1995) analyzes contemporary cities, focusing in particular on the conditions for connections, sometimes distant and not very apparent, between centers and peripheries. He deciphers the influence that these dynamics may or may not have on the emergence of a new form of interdependence, multipolarity and functional links in everyday life, a form that he terms a *metapole*: "A set of spaces in which all or part of the inhabitants, economic activities or territories are integrated into the daily (ordinary) functioning of a metropolis" (Ascher 1995, p. 34, *author's translation*).

In this chapter, we propose to revisit this query on the connection of metropolises with their periphery, in the context of environmental mutations, climate change and associated evolutions of risk management doctrine. Ascher examined the connections made possible through technology or the evolution of social and economic practices. In this chapter, we propose to consider the connections that have become salient by making visible the existence of a strong hydrological link between various territorial units.

Our working hypothesis is that contemporary flood risk management creates new "meta-political" interrelations and new ways of judging the relationship between an urban center and its connected periphery. Today, flood risk management is no longer solely based on protection infrastructures, but rather on a set of territorial governance tools intended, on

---

Chapter written by Jean-Paul VANDERLINDEN and Nabil TOUILI.

the one hand, to limit the exposure of goods and people affected by a spatial displacement of the effects of the hazard and, on the other hand, to ensure that information on the risk of flooding is shared with the people affected. The doctrine of flood risk management in France has gone through different phases whose successive and often partial implementations have generated a jumble of structural and non-structural measures (for further reading, see Gilbert and Gouy (1998); Bayet (2000); Barraqué and Gressent (2004); Anselme et al. (2008); Erdlenbruch et al. (2009); Deboudt (2010); Lumbroso et al. (2011); Hissel et al. (2015)).

Storm Xynthia in 2010 renewed the impetus for the implementation of territorialized approaches. The associated acceleration, highlighted by many actors (Touili and Vanderlinden 2017), was accompanied by an extension of the areas taken into account in risk management processes. Today, risk governance thus operates at all scales through a series of instruments aiming at an important integration not necessarily linked to administrative scales and delimitations. In such a context, solidarity in the face of flood risk takes the form of regional solidarity on the scale of enlarged basins where interventions can be modulated according to local, hydrogeological, social or economic characteristics. The governance of risks thus takes a form that is more respectful of hydrological connections, remote, making visible forgotten interrelations between the center and periphery of metapoles.

In this chapter, we question how these reconnections are implemented, and in particular what frameworks are mobilized by actors to make sense of this evolution. We focus on flood risk governance as a factor that reinforces the integration of metapoles. This governance of flood risk is now part of the "daily (ordinary) functioning" (Ascher 1995, p. 34, *author's translation*) of central cities protected by their peripheries. Land that is less valuable, or whose owners are less influential, is perceived or designated as "sacrificed" to protect the dense city. This leads to trade-offs, for which the values of the actors involved are mobilized. We therefore turn our attention to a reading of the ethical dimensions associated with the very process of metapolization.

5.1. The conceptual framework of narrative analysis

Our corpus is approached from the perspective of narrative analysis (Josselson 2011) with the aim of identifying a series of meta-narratives associated with different ethics, which can be convened in order to make sense of the evolution of risk governance. This approach opens privileged access to the question of the link between flood risk, sensemaking and the context of metapolization.

The first step is to collect the narratives of the actors interviewed. Next, we identified in these narratives the statements expressing a moral judgment on the situation. We associated these judgments with the ethical theories they invoke. We then grouped these different narratives and statements in the form of typical meta-narratives, each meta-narrative being associated with an ethical theory. Finally, we analyzed how these meta-narratives make sense of the changes observed.

5.1.1. *Stories of risk governance*

In the context of studying climate risk, here closely related to flood risk, narratives are generally defined as accounts of events or processes with temporal or causal coherence (IPCC 2014, p. 202). Narratives, or stories, are dynamic and dialogical, reveal a set of values and meanings, can take a variety of forms, carry individual identities and serve as a medium for sharing experience, values, that ground a collective identity (Rommetveit et al. 2013; Fløttum and Gjerstad 2017; Moezzi et al. 2017; Krauß 2020; Krauß and Bremer 2020).

Regardless of their importance and their capacity to encapsulate human experience in all its diversity, narratives inform us through the existence of meta-narratives. These relate more to the existence of collectives, shared visions of the world, of that which governs it or should govern it. These meta-narratives constitute "larger explanations of our reality that guide our own smaller narratives. Meta-narratives explain in an overarching fashion why we do what we do and thus define our view of the world or part of it" (Badke 2012, p. 104, *author's translation*).

The hypothesis that narratives are carried by one or more meta-narratives centrally convoking the values of speakers is now regularly validated (Kahan et al. 2012), including in the works we have conducted in Gironde (Touili et al. 2014; Touili and Vanderlinden 2017; Vanderlinden et al. 2017). This premise, following on from the seminal work of Douglas and Wildawski (1983), is based on the treatment of risk as a shared cultural entity, albeit at varying scales and degrees.

5.1.2. *Sensemaking as a source of narratives about change*

When confronted with changes, individuals and human communities place these into a narrative. As groups, they make individual and collective sense of these changes in light of their own knowledge, beliefs and experiences (Ketelaar et al. 2012). The process of putting changes in the institutional and natural environment into narratives is referred to as sensemaking. Sensemaking is defined as the process of interaction between the frame of reference of individuals and collectives confronted with change, and their perception of the demands inherent in these changes (Luttenberg et al. 2013). In a situation of risk and the evolution of risk governance, as outlined in the previous section, such a process of sensemaking implies, among other things, the convocation of a particular category of frame of reference: ethical theories, both individual and collective, serving as a theoretical framework for the judgments expressed. In this case, the narratives collected and analyzed are those of actors articulating the evolution of flood risk management methods.

Sensemaking in risk governance situations involves drawing from a library of interpretive frameworks. Among these interpretive frameworks, ethical theories are prominent (Douglas and Wildawski 1983; Kahan et al. 2012; Touili et al. 2014; Touili and Vanderlinden 2017; Vanderlinden et al. 2017).

5.1.3. *A corpus of interviews on the risk of flooding in Gironde*

We applied this narrative reading to a corpus of 33 interviews conducted with various institutional and non-institutional actors in flood risk management in the Gironde department. This corpus has the significance of having been collected while the effects of the 2010 storm Xynthia were still very much a fresh memory. The evolution of the flood risk management framework was the subject of dynamic debate. The hydraulic interconnections between Bordeaux and its periphery were visible and occupied many minds.

In this corpus, through thematic coding, we have isolated a series of statements that express the speakers' values. The identification of these statements is based on the presence of causalities related to floods or their management, as well as on the presence of judgments regarding these causalities and the actions of the actors, whether these judgments are stated explicitly or implicitly.

When a statement expresses a judgment about a situation, it is a statement that calls upon the speaker's ethics, that is, their operational capacity to distinguish, by mobilizing their values, between what is desired and what is not desired. This operational capacity to mobilize values is based on a framework identifying, in a general and transposable way, the rules that apply and the conditions of their application. This framework is the ethical theory upon which a judgment is based. The expression of a judgment is therefore doubly informative: information about what the speaker considers desirable, and information about what substantiates this judgment.

Ethical theories can be categorized according to the relative importance given to (a) the structure of the situation, (b) the intrinsic qualities of the agent involved, or (c) the consequences observed empirically. If the emphasis is placed on the structure of the situation as such, before noting the consequences linked to this situation, we speak of *deontology*. If the focus is on a specific quality of the agent (benevolence, honesty and other notions associated with virtuous individuals), we speak of *virtue ethics*. Finally, if the focus is on the evaluation of consequences, we speak of *consequentialism*.

For the analysis of our interview portfolio, we have linked the collected statements to these three categories of predetermined ethical theories. To these categories, we added categories that emerged along the way (justice, equity, distribution, deliberation ethics, ethics of nature).

These analyses place us in the presence of statements that express a judgment, statements that we have grouped according to the ethical theories they share. We then associate these ethical theories with the meta-narratives that the identified statements convoke. These steps allow us to identify the resources in terms of ethical theories that the actors call upon to make sense of the changes they observe, and to associate the meaning that these actors give to the situation they observe.

5.2. Ethical theories invoked and associated meta-narratives

The analysis of the corpus has allowed the identification of a series of meta-narratives associated with different ethical theories: deontology manifests itself in the context of debates around the strict application of rules (this is indeed a highly regulated environment); virtue ethics manifests itself via different statements related to issues of ex ante justice; consequentialist ethics manifests itself in a pragmatic form (it is about reducing concrete risks on goods and people), as well as in terms of ex post territorial equity. Finally, we have observed an interest in a deliberation ethics and in the adoption of an ethics of nature.

5.2.1. *Deontology in terms of having respect for shared norms: the meta-narrative of deontological hype*

A first ethical theory that has guided our analysis is that of deontology in terms of having respect for formalized rules (laws and associated official texts) that are imposed on the collective. It is a question of narratives aimed at laws, rules, regulations and the respect for these is imposed on everyone. These narratives use rules and public policies, either as self-imposed by their normative nature, or as a starting point for alternative norms.

As we pointed out in the introduction, flood risk management today is characterized by numerous political and regulatory injunctions. It is quite natural that the question of compliance arises:

> What we can do, as a state department, is to ferociously, unambiguously, unequivocally impose bans (Departmental Risk Manager[1], speaking on the links between the climate context and the regulatory context, *author's translation*).

Nevertheless, these narratives that invoke the rules also emphasize their potentially deleterious effects: when it comes to application, "ferocity" is not always well-received; the letter and the spirit of the text must be distinguishable, the latter sometimes having to prevail:

> It is rather in the opposite direction that we must reason, without necessarily looking at the regulatory document, [it is

---

1 *Responsable départemental gestion des risques.*

preferable to] think about the safety aspect, the aspect of economic damage that a flood could generate (Intercommunal Risks and Nuisances Executive[2] speaking about the issues of land use planning in relation to risk management, *author's translation*).

Since storm Xynthia, stricter application of the regulatory texts has been the subject of renewed attention, sometimes with strong enthusiasm:

The only observation that everyone makes is that today [after storm Xynthia], we need to [...] go further in risk prevention with, in my opinion, strong prohibition measures (Port Infrastructure Manager[3], speaking on recent developments within his operating environment, *author's translation*).

However, this voluntarism is not shared systematically:

We don't know how to do the right thing or when to balance: we do nothing or almost nothing, or else too much [...] And we've ended up doing too much and badly (Member speaking on behalf of a citizens' association, about the interactions between land use planning and flood risk management, *author's translation*).

The statements identified as invoking an ethical theory of compliance shares a meta-narrative that we have termed "deontological hype". This meta-narrative is structured along the following lines:

– flood risk management is a necessity in the Gironde estuary;

– in the Gironde, we are faced with numerous regulatory texts and practices intended to manage the flooding risk;

– until 2010 (storm Xynthia), their application was rather flexible and adapted to the different specificities of local contexts;

– storm Xynthia marked a break in flood risk management and a rush in the application of regulations;

---

2 *Cadre intercommunalité, risques et nuisances.*
3 *Gestionnaire d'infrastructures portuaires.*

– since then, strict and inflexible application of the rules creates harmful situations.

We observe here that the sensemaking associated with the post-storm Xynthia reinforcement of the regulatory frameworks is not limited to the invocation of human and material losses associated with storm Xynthia. The governance of flood risk, from which sense is being made, is a question of balance, of respect for local variations and the various needs of actors in their diversity. This observation foreshadows the rest of our results.

5.2.2. *Virtue ethics: the meta-narrative of "respect for justice" as a virtue*

The second ethical theory that guided our analysis is that of virtue ethics, touching on questions of the intrinsic qualities of the agents ensuring the governance of flood risk: Do the decisions envisaged demonstrate benevolence, generosity, empathy?

In this context, we observe two indicators of virtue. The first indicator, which is related to the issue of justice, consists of refusing to consider situations that degrade those of certain actors through the implementation of risk management measures designed to protect other actors. This is an ex ante observation: it is not a question of observing a result, but of considering the virtue (or lack of virtue) of an agent contemplating a unilateral distributive action. In this framework, an agent who refuses such a measure is considered virtuous; a contrario, another who considers or promotes the measure is judged negatively.

> One of the principles is to say, in order to relieve the areas at stake: "We're going to recreate flood zones [elsewhere]," and that's not logical [meaning here acceptable]. And the local elected officials [...] then say, "You don't want to protect us, you want to flood us to protect Bordeaux" (Departmental Official[4], risk management, *author's translation*).

This indicator of the existence of a virtue ethic is also linked to the expression of an injunction of ex ante compensation. In order to correct situations where the implementation of a measure is deemed impossible as a

---

4 *Responsable départemental.*

result of the feeling of injustice generated, it is possible to provide for compensatory measures, consubstantial with the measure taken, aimed at changing the net effect, ex ante, of the measure envisaged:

> [for some] areas that would be flood-prone and for which we would want to put […] in an effort, […], we would [compensate] (Departmental Official, agriculture and related, *author's translation*).

These statements associated with virtue ethics are structured by a meta-narrative that we call "respect for justice as a virtue", which is structured as follows:

– flood risk management is a necessity in the Gironde estuary;

– the selection of a risk management option and its implementation are conditional on the absence of inequitable or perceived inequitable distributional effects;

– if distributional effects penalizing a group are anticipated and not corrected, the measure envisaged must be rejected so as not to create a prejudicial situation.

When virtue ethics is invoked, actors engage in sensemaking by interpreting the changes underway as potentially generating unacceptable distributive impacts. In this framework, transition periods are periods of potential wealth transfer, in the broad sense, and monitoring must be carried out to mitigate against any inequitable transfers due to the possible presence of non-virtuous agents. Once again, we observe the creation of sensemaking that emphasizes the heterogeneity of individual and collective situations, and the importance of taking them into account.

5.2.3. *Consequentialism in risk reduction: the meta-narrative of ordinary risk governance*

A series of statements focused on the ex post consequences of adopted governance measures are identifiable in our portfolio of interviews. These citations are associated with the obvious hydrological dimensions of exposure to flooding:

If you want to preserve the populated areas, where there is the most at stake, you have to send the water to the empty spaces. That's a technical reality (Departmental Land Management Executive[5], speaking about the options for land transformation under the pressure of evolving risks, *author's translation*).

For this group of people, the rules must make it possible to prevent flooding of something that is valued by humans. If they fail to do so, then the consequences of these encounters must be limited. This is a standard way of assessing flood risk management in terms of consequence reduction. We group these statements under the meta-narrative of "ordinary risk governance", which is structured as follows:

– flood risk management is a necessity in the Gironde estuary;

– it is a set of concrete operations that have an effect on the territory;

– it is possible that the effects are not those expected, both in terms of efficiency and in terms of induced effects;

– these risk management operations must be evaluated in terms of their capacity to reduce the risk to which people and properties are exposed.

The analysis of statements about the consequences of flooding and the judgment of management measures in terms of hydraulics and avoidance indicates that sensemaking is at work. The meaning created is linked to the existence of a culture of flood risk. Stakeholders make sense of observed changes by invoking their past observations in a flood-prone environment.

5.2.4. *Consequentialism in terms of inequity: the meta-narrative of the questionable fairness of choices made*

Within our portfolio of interviews, it is possible to identify statements very close to those already observed in the meta-narrative of virtue ethics. These are statements that question decisions not in terms of "what is to come" (ex ante), but in terms of what has taken place and which has an observed effect (ex post). In the meta-narrative of "respect for justice as a virtue", the rigidification of rules following storm Xynthia is criticized because of the unjust situation, or the perception of it as such, that it creates

---

5 *Cadre direction départementale du territoire.*

ex ante. For some, the consequences are already being felt in their daily lives. What is in question is not the expression of a structural deficit of justice, but rather the ex post observation of new constraints deemed unfair:

> We have farmers who [...] [do not accept] an accentuation of the risks linked to the programmed disappearance of dikes on agricultural land to protect the Bordeaux agglomeration – an agglomeration that has spread over flood prone areas (Engineer in charge of studies, research center, speaking about the issues of land use planning in relation to risk management, *author's translation*).

Statements about the inequitable outcome, ex post, of decisions made are underpinned by a meta-narrative that we call "the questionable fairness of choices made":

– flood risk management is a necessity in the Gironde estuary;

– it is a set of concrete development operations that have an effect on the territory;

– it is possible that some consequences are not those expected, both in terms of effectiveness and in terms of induced effects;

– these risk management operations are evaluated in terms of the equity of the distributive impact that they may have, particularly in terms of securing a particular territory.

The associated sensemaking does not question the necessity of flood risk management; it offers an interpretative framework that resonates with that associated with virtue ethics; proactive monitoring is necessary and the inventory, along the way, of positive and negative effects is required. It is not so much a question of debating the often-contradictory nature of the search for efficiency and justice, but rather of empirically supporting this debate in a way that is deemed satisfactory.

5.2.5. *Deliberation ethics: the meta-narrative of the process that is to be improved*

Some of the narratives in our portfolio of interviews evoke a deliberation ethics (Habermas 1991) that seems to have a fairly clear demand:

> The development of the territory is imposed by the State without any consultation. In spite of a mini-concertation, we receive people from the DDTM[6] [Departmental Directorate of Territories and the Sea], in charge of civil protection. We receive them, but they present us with a *fait accompli* (Engineer in charge of studies, research center, speaking about the issues of land use planning in relation to risk management, *author's translation*).

It is not so much the lack of a formal procedure as the limitations associated with the procedures used, for example:

> There are some who believe that anything that is discussed is accepted. It's not quite that simple. The culture of consultation is not yet very well integrated in our country (Risk Manager, Departmental Directorate for the Territory[7], *author's translation*).

We associate these statements with a meta-narrative that is related to those associated with deontology, virtue ethics and consequentialism:

– flood risk management is a necessity in the Gironde estuary;

– it is a set of concrete development operations that have an effect on the territory;

– it is possible that the effects are not those expected, both in terms of effectiveness and in terms of induced effects;

– these operations must be supported by all the parties involved and demonstrate efficiency, justice and equity;

– fulfilling these conditions is only possible through the organization of informed deliberations with all the parties involved.

In a decision-making context, this is a request for the adoption of deliberative procedures. The purpose is no longer simply to guide the decision but to establish its procedural quality. From the point of view of the

---

6 *Direction départementale des territoires et de la mer.*
7 *Direction départementale du territoire.*

sensemaking, the invocation of deliberation ethics indicates that the changes observed by the actors are interpreted in terms of a lack of involvement of the stakeholders concerned. The elements judged negatively in other respects – efficiency, justice, equity – can be interpreted via this deficit.

5.2.6. *Ethics of nature: the meta-narrative of nature holds the keys*

Finally, we identify in the portfolio of interviews the expression of an ethics of nature. These are statements that evaluate situations in terms of the tension that can exist between nature, its functioning and human practices:

> Why not restore the natural character of these diked areas, let nature finally reclaim its rights […] and let the animals, the little creatures come and settle in this territory? (R&D Manager, State Department, risk management, considering future paths, *author's translation*).

> Nature will win, anyway (Departmental Manager, risk management, identifying areas at risk and explaining their nature and the nature of the risk, *author's translation*).

We group the statements invoking nature as a source of explanation, as a model serving as inspiration, under the meta-narrative that we call "nature holds the keys" and whose structure is the following:

– flood risk management is a necessity in the Gironde estuary;

– the challenges of this governance are related to the illusion that we can shape our environment as we wish;

– the connections, sometimes not very visible and distant, are fundamental;

– our practices and organizations would benefit from taking into account these so-called natural dimensions.

The invocation of an ethic of nature indicates sensemaking centered on our relationship to the *other*, this time non-human. Our failures lie in the nature–culture dichotomy which, in the case of flood risk management, blinds us.

5.3. For deliberative risk governance

The results of our analysis inform us about two fundamental dimensions that will be discussed. On the one hand, they show the entangled coexistence of ethical theories generating a variety of sensemaking dynamics in the face of changing risk governance doctrine. On the other hand, these results show that the actors wish to question the relationship to the *other* as actors involved in this governance, and even to the *other* as nature, the matrix of the issues associated with this risk governance.

The ethical theories mobilized in the collected accounts do not therefore fall into a monolithic category. The actors narrate situations to which they associate a judgment of risk and its mitigation which, depending on the particular situation, is:

– the deontology of respect for the rules;

– virtue ethics;

– consequentialist ethics;

– a combination of two or three ethical theories.

Moreover, for the same management option, the invocation of these ethical theories can lead to opposing judgments. Risk, like the uncertainties that are consubstantial with it, has the characteristic of generating ambiguities through the multiplicity of meanings created. These ambiguities impose arbitrations that are structurally related to moral values. Furthermore, the invocation of a diversity of ethical theories to create meaning in a situation of change means that the inter-interpretation of the observed change can generate situations where the same event is interpreted, and put into meaning, in mutually incompatible ways.

These results resonate with those presented by Ersdal and Aven (2008), who analyze how decision-making in situations of uncertainty, depending on the decision frameworks chosen, can call upon deontology (see "deontological hype" in our results), virtue ethics (see "respect for justice as a virtue" in our results) and judgments based on consequences (see "ordinary risk governance" in our results). Ersdal and Aven (2008) link the dominant approaches to decision-making, each individually, to different ethical theories, for example: cost–benefit analysis and consequentialism, acceptable risk threshold and precautionary principle under virtue ethics.

Ersdal and Aven (2008) also show that the two frames of reference can be and are combined in the context of managerial approaches to risk management. It is a similar result, obtained at higher resolution, that our results point to. The entanglement of ethical theories requires that flood risk governance be thought of in a way that respectfully takes this diversity into account. This type of situation leads various authors (Renn 2008; Kahan et al. 2012; Vanderlinden et al. 2017) to advocate for the establishment of forums that promote deliberation, thereby ensuring all stakeholders are heard. The idea is to advocate for a mechanism that allows for the diversity of ethical theories invoked to contribute to decision-making.

Moreover, the global changes facing the Bordeaux metropolis call for significant adaptive transformations. Exploring the links between mobilized ethical theories and adaptive transformation, Adjibade and Adams (2019) propose a series of recommendations that resonate with our own results. Four of these recommendations fall directly under deliberation ethics:

– practice participatory foresight and decision-making;

– create conditions where the balance of power is balanced;

– create and nurture spaces for experimentation and mutual learning;

– foster flexible, decentralized and adaptive governance.

One of these recommendations is directly related to virtue ethics: avoid the redistribution of risk. Our results therefore point not only to demands for flood risk governance, but also to demands that resonate with a situation of adaptation to climate change.

In the case of the Gironde meta-political territory, the interests involved are multiple and contradictory. A network of actors linked by hydrology, nature and regulation is associated with the daily management of risk in Gironde. This brings us back to the starting point of our exploration – that of the visibility of a metapolization via hydrology and flood risk governance; in a context where the risk of flooding will increase, whether it be exceptional floods or marine submersions due to an increase in the intensity of storm surges (Oppenheimer et al. 2019), in the Gironde (Hissel et al. 2015).

This visibility, through sensemaking, and the associated convocation of ethical theories by actors, underlines that a metapole is a whole characterized by diversity that potentially generates two central demands in the governance of climate change: deliberative inclusive governance and the consideration of nature. These demands are not just ends within themselves. They can also be conceived as a means of monitoring the effectiveness of risk governance, including justice criteria (ex ante) and equity (ex post) – following the sensemaking associated with deontological, virtuous and consequentialist ethical theories.

5.4. Conclusion

The results presented and the associated discussion allow us to return to our initial question. Faced with the diversity of sensemaking at work, how can we think about metapolization? Let us return to Ascher's book *Metapolis, or the Future of Cities* (1995), and to one of the conclusions that Sander and Vergès (1996) drew at the time:

> The absence of a metapolitan institution could be an opportunity to promote an ad hoc study structure, such as that of the master plan, as a permanent and renewed management structure for a "city project" that could become a city project, in a process that relies on all of the actors involved, as early as possible (*author's translation*).

In 1995, metapolization already called for governance that invoked the ethics of discussion. Now, it seems important to us to consider adding the practical development of an ethic that takes into account our relationship with nature and the effect of this relationship, both on us and on nature. This will equip us to face global changes in their impact on the Bordeaux metropolis. The plural nature of the meanings created in a situation of change implies adopting an attitude that allows us to take into account the associated pluralism.

While there was a time when governance could be organized around economic and social realities, it is now desirable, in these times of environmental change and ecological transition, for us to collectively accept that ecological and hydraulic realities remind us of our ethical positions. Let us then mobilize ethical theories in order to negotiate a shared meaning.

5.5. References

Adjibade, I. and Adams, E.A. (2019). Planning principles and assessment of transformational adaptation: Towards a refined ethical approach. *Climate and Development*, 11(10), 850–862.

Anselme, B., Goeldner-Gianella, L., Durand, P. (2008). Le risque de submersion dans le système lagunaire de La Palme (Languedoc, France) : nature de l'aléa et perception du risque. In *Actes du colloque international pluridisciplinaire. Le littoral : subir, dire, agir*. Lille.

Ascher, F. (1995). *Métapolis ou l'avenir des villes*. Odile Jacob, Paris.

Badke, W.B. (2012). *Teaching Research Processes. The Faculty Role in the Development of Skilled Student Researchers*. Chandos Publishing Group, Hull.

Barraqué, B. and Gressent, P. (2004). La politique de prévention du risque d'inondation en France et en Angleterre : de l'action publique normative à la gestion intégrée. Report, Ministère de l'Écologie et du Développement durable, École nationale des ponts et chaussées, Université de Marne-la-Vallée et Université Paris XII, Paris.

Bayet, C. (2000). Comment mettre le risque en cartes ? L'évolution de l'articulation entre science et politique dans la cartographie des risques naturels. *Politix. Revue des sciences sociales du politique*, 13(50), 129–150.

Deboudt, P. (2010). Vers la mise en oeuvre d'une action collective pour gérer les risques naturels littoraux en France métropolitaine [Online]. Available at: https://doi.org/10.4000/cybergeo.22964.

Douglas, M. and Wildawski, A. (1983). *Risk and Culture: An Essay on the Selection of Technological and Environmental Dangers*. University of California Press, Berkeley.

Erdlenbruch, K., Thoyer, S., Grelot, F., Kast, R., Enjolras, G. (2009). Risk-sharing policies in the context of the French flood prevention action programmes. *Journal of Environmental Management*, 91(2), 363–369.

Ersdal, G. and Aven, T. (2008). Risk informed decision-making and its ethical basis. *Reliability, Engineering & System Safety*, 93, 197–207.

Fløttum, K. and Gjerstad, Ø. (2017). Narratives in climate change discourse. *Wiley Interdisciplinary Reviews: Climate Change*, 8(1), e429.

Gilbert, C. and Gouy, C. (1998). Flood management in France. In *Flood Response and Crisis Management in Western Europe*, Rosenthal, U. and Hart, P. (ed.). Springer, Berlin/Heidelberg.

Habermas, J. (1991). *De l'éthique de la discussion*. Éditions du Cerf, Paris.

Hissel, F., Baztan, J., Bichot, A., Brivois, O., Felts, D., Heurtefeux, H., Vanderlinden, J.-P. (2015). Managing risk in a large flood system, the Gironde estuary, France. In *Coastal Risk Management in a Changing Climate*, Zanuttigh, B., Nichols, R., Vanderlinden, J.-P., Burcharth, H.F., Thomson R.C. (eds). Elsevier/Butterworth-Heinemann, London.

IPCC (2014). Climate change 2014. Impacts adaptation, vulnerability. Part A: Global sectoral aspects. Report, Intergovernmental Panel on Climate Change, Geneva.

Josselson, R. (2011). Narrative research. Constructing, deconstructing, and reconstructing story. In *Five Ways of Doing Qualitative Analysis. Phenomenological Psychology, Grounded Theory, Discourse Analysis, Narrative Research, and Intuitive Inquiry*, Wertz, F.J., Charmaz, K., McMullen, L.M., Josselson, R., Anderson, R., McSpadden, E. (eds). Guilford Press, New York.

Kahan, D.M., Peters, E., Wittlin, M., Slovic, P., Larrimore Ouellette, L., Braman, D., Mandel, G. (2012). The polarizing impact of science literacy and numeracy on perceived climate change risks. *Nature Climate Change*, 2(10), 732–735.

Ketelaar, E., Beijaard, D., Boshuizen, H.P., Den Brok, P.J. (2012). Teachers' positioning towards an educational innovation in the light of ownership, sense-making and agency. *Teaching and Teacher Education*, 28(2), 273–282.

Krauß, W. (2020). Narratives of change and the co-development of climate services for action. *Climate Risk Management*, 28, 100217.

Krauß, W. and Bremer, S. (2020). The role of place-based narratives of change in climate risk governance. *Climate Risk Management*, 28, 100221.

Lumbroso, D., Stone, K., Vinet, F. (2011). An assessment of flood emergency plans in England and Wales, France and the Netherlands. *Natural Hazards*, 58(1), 341–363.

Luttenberg, J., van Veen, K., Imants, J. (2013). Looking for cohesion: The role of search for meaning in the interaction between teacher and reform. *Research Papers in Education*, 28(3), 289–308.

Moezzi, M., Janda, K.B., Rotmann, S. (2017). Using stories, narratives, and storytelling in energy and climate change research. *Energy Research & Social Science*, 31, 1–10.

Oppenheimer, M., Glavovic, B.C., Hinkel, J., van de Wal, R., Magnan, A.K., Abd-Elgawad, A., Sebesvari, Z. (2019). Sea level rise and implications for low-lying islands, coasts and communities. Report. IPCC, Geneva.

Renn, O. (2008). *Risk Governance: Coping with Uncertainty in a Complex World.* Earthscan, London.

Rommetveit, K., Gunnarsdóttir, K., Jepsen, K.S., Bertelsson, M., Verax, F., Strand, R. (2013). The TECHNOLIFE project: An experimental approach to new ethical frameworks for emerging science and technology. *International Journal of Sustainable Development*, 16(1/2), 23–45.

Sander, A. and Vergès, V. (1996). Notes de lecture, Métapolis ou l'avenir des villes. In *Métropolisation : interdépendances mondiales et implications lémaniques*, Leresche, J.-P., Joye, D., Bassand, M. (eds). Georg, Geneva.

Shrader-Frechette, K.S. (1991). *Risk and Rationality. Philosophical Foundations for Populist Reforms.* University of California Press, Berkeley.

Touili, N. and Vanderlinden, J.-P. (2017). Flexibilité adaptative et gestion du risque : étude de cas des inondations dans l'estuaire de la Gironde. *Vertigo*, 17(2).

Touili, N., Baztan, J., Vanderlinden, J.-P., Kane, I.O., Diaz-Simal, P., Pietrantoni, L. (2014). Public perception of engineering-based coastal flooding and erosion risk mitigation options: Lessons from three European coastal settings. *Coastal Engineering*, 87, 205–209.

Vanderlinden, J.-P., Baztan, J., Touili, N., Kane, I.O., Rulleau, B., Simal, P.D., Zagonari, F. (2017). Coastal flooding, uncertainty and climate change: Science as a solution to (mis)perceptions? – A qualitative enquiry in three European coastal settings. *Journal of Coastal Research*, 77(sp1), 127–133.

PART 3

When the Estuary Makes the City

6

The Estuarine City as an Allegory for Changes in Solidarity

The estuarine and coastal city has an allegorical value in terms of the transformations at work in ways of unified thinking, that is, solidarity. In coastal areas, the tensions between economic attractiveness and ecological and social vulnerability can be observed in a spectacularly radical way[1]. However, there are also natural resources (MPAs, sentinel forests, wetlands, etc.)[2], even if they have been neglected for a long time, to counter the effects (de Godoy Leski et al. 2019; de Godoy Leski 2021). The notion of allegory involves speaking about an abstract idea through the use of imagery and historical narrative. In this case, the idea concerns the government of a complex environment with both ecological and socio-political meaning. The context is based on the history of Bordeaux and the Gironde estuary.

A major change in the government of French society is the assignment to cities of functions that the State is currently struggling to perform, not only in terms of territorial planning and sustainable development, but also in terms of social cohesion. Emmanuel Macron's programmatic work, *Révolution* (2016), insists on the "great responsibility" of each city "with regard to the territory in which it is located" (Macron 2017 [2016], p. 154,

---

Chapter written by Thierry OBLET.

1 The risk of flooding, associated with rising sea levels during storms, makes Aquitaine particularly exposed to the effects of global warming.
2 According to biological oceanography specialists, marine protected areas (MPAs) can mitigate the impact of climate change and contribute to the adaptation of ecosystems and populations.

author's translation). The estuarine city illustrates the evolution of the notions of interdependence and solidarity. It constitutes a privileged field of observation in terms of the challenges associated with the transition from a 20th century, solidarity-based framework of state governance that recognizes the interdependence of individuals within the nation, to a 21st century, solidarity-linked framework of metropolitan governance that recognizes the interdependence of territories. It should be noted at the outset that this second conception of solidarity is not intended to replace the first, but to respond to problems that the latter is facing.

This shift is not an abrupt break. If the demand for territorial equality is presented today as a response to the threat to social cohesion posed by the predatory dimension of cities (Faburel 2018), it also appears to be the logical extension for the recognition of human interdependence under the Third Republic. Much more than a scientific constant, the proclaimed equality of territories re-conduces that act of political faith that was already contained in the idea that all people are born free and have equal rights (Dubet 2019). However, neither the State nor the city is a univocal figure. This is why the estuarine city has a specific symbolic dimension, and the historical narrative proposed here, based on the Girondin case, aims to put two lessons into perspective.

In terms of the logic that presides over the globalized economy, metropolization is akin to "a deployment of complex forms of knowledge and creativity that are too often accompanied, aside from solid profits, by extraordinarily primary brutalities" (Sassen 2016, p. 293, *author's translation*). The first symbolic challenge of the estuarine city is to detach the metropolitan idea from a demonization induced by its assimilation to a product of neoliberal globalization marked by exacerbated consumerism (*1 – Cleaning the metropolitan idea of the stench of its emissions and ecological irresponsibility*).

What is the meaning of this rather cryptic formula of territorial equality, which seemed, during the five-year term of François Hollande, to constitute the moral framework for the recognition of the interdependence of territories? This egalitarian claim could have been seen as an anachronistic attachment to uniformity, when the efficiency of public policies required knowing how to distinguish between the drivers of change and the wheels of progress. In the era of cities, this means ensuring that the equality of territories does not thwart metropolitan innovation, a source of economic growth (Davezies 2012). In an assumed metropolization, claiming territorial equality is a long-term way of reaffirming the primacy of the territory over the domain within the institutional register. This operation has been associated in history with the promise of a fairer and better-informed sovereignty (*2 – From the conquest of domains to the recognition of equality for territories*).

For the past 10 years or so, elected city officials, particularly those in Bordeaux, have tended to recognize that a large city needs to base its development on efficient secondary poles. However, this attention appears to be wishful thinking because, viewed solely in economic and social terms, the interdependence between the resources of the city and those of the suburbs appears to be too unbalanced in favor of the center. The strength of the Yellow Vest Movement in Gironde has confirmed this (Vermeren 2019). The second symbolic issue of the estuarine city is to accredit the idea of a rebalancing of relations between the center and the periphery. The inclusion of natural resources (water, land, food, ecosystem services, etc.), in addition to the usual economic and social assets considered from the perspective of the tensions between the metropolization of jobs and the peri-urbanization of housing, seems likely to provoke a reassessment of the perceived usefulness of the various territories of Gironde in the production of the common good (*3 – From equality to the cohesion of territories*).

6.1. Cleansing the metropolitan idea of the stench of its emissions and ecological irresponsibility

With the emergence of industrial society, the idea that men are defined more by their activities than by their birth was imposed. The role of sociology was to make individuals aware that the recognition of their complementarity produced by social specialization formed the basis of their integration and the foundation of the republican regime. In the prism of this rational and moral representation, Émile Durkheim was able to reassure his contemporaries that they need not worry too much about their growing dissimilarity in beliefs and ways of life. Civilization was not collapsing; it was being transformed (Durkheim 1898).

However, public policy had to take steps to heighten awareness of this solidarity, based on a sense of mutual usefulness rather than on the similarity of morals. This required compensating for the unequal exposure of workers to the risks of work (accidents, illness, unemployment). The creation of social insurance was both the means and the consequence of this awareness. This construction of a social state did not come about naturally. Elites had to work to theorize this "organic solidarity", which abolished neither social hierarchies nor the reasons for conflicts between employers and workers. This work found a political relay in solidarism. It was certainly stimulated by social struggles, but it also gave them a horizon in return: the development of social rights (Donzelot 1984).

In addition to the development of individual social rights, the social state, in its project to concretize this solidarity, had, in the 20th century, shown concern for homogenizing the territories with the goal to equalize access to public services. Prefectures, sub-prefectures and canton capitals served all the more as a framework for its deployment since France, contrary to the English model of the large coal-mining city, had become industrialized around its small- and medium-sized cities, which were the first destinations for peasants abandoning their land (Desjardins and Estèbe 2019).

This dream of uniformly offered services, in each department[3], symbolized the idea of territorial equality, which in France, due to its very different urbanization, made its organization particularly costly (Estèbe 2015). The polarization of economic growth in certain cities was an accepted fact, and it was the responsibility of regional planning policies to ensure that the modernization of the country took place in a balanced manner. The "cities of equilibrium" created in the mid-1960s[4] were supposed to compensate for the influence of Paris through their capacity to provoke economic growth within their region. The large cities financed rural areas through a redistribution regulated by the State, where "national" policies, targeting individuals rather than territories (social benefits and pensions, non-market public service wages), played a decisive role (Davezies 2008).

This pattern of development ran out of steam with the shift, at the turn of the century, from an industrial society to a "hyper-industrial" society (Veltz 2017), merging industry and service in the context of the digital revolution, and enhancing competition between cities. The issue is no longer the submissive attachment of populations to a "factory for life", but rather their flexibility and mobility within the context of a globalized economy where

---
3 With standards that are often not adapted to local idiosyncrasies (Oblet 2005).
4 The urban community of Bordeaux dates from 1968.

information processing offers greater added value than the transformation of matter. Large cities become economic actors in their own right as soon as they succeed in acquiring, through a project, a maximum number of metropolitan attributes: command center, resource center, crossroads of multiple communication networks, showcasing a certain art of living (Oblet 2005).

Despite a death foretold 40 years ago (Vingré 1980), the social state has not disappeared. However, for those who do not neglect the imaginary dimension of human works and actions, the metropolitan boom can be associated with the transition from a myth of the state, to a myth of the city. The myth of the state was constructed at the convergence of the Enlightenment and the French Revolution. It symbolized the belief in the possibility of transforming society through appropriate public policies and education (Baczko 1988).

The myth of the city valorizes the idea that cities are better positioned than states to address the major social, cultural, economic and ecological challenges of the 21st century. Benjamin R. Barber's highly institutionalized book, *What If Mayors Ruled the World* (2015), is just one illustration of this among many. Another is the association *Notre affaire à tous* ("It Concerns Us All"). Instigator of the petition *L'Affaire du siècle* ("The Story of the Century"), its manifesto for climate justice insists on the importance of cities in the legal and financial battles waged against toxic multinational corporations (*Notre affaire à tous*, pp. 22–26, *author's translation*). Finally, is it not significant that the policy in charge of restoring social cohesion, where it seemed most threatened, has taken the name "city policy" (Donzelot 2006)?

The city of equilibrium emanated from state-directed urban planning. The "global city" at the end of the 20th century illustrates a process of metropolization emancipated from its tutelage. Admittedly, the notion of the "global city" (Sassen 1996) was only appropriate for a small number of cities integrated into the exchanges of the globalized economy, those capable of focusing their development on finance and high technology; nevertheless, it did outline a general orientation, that of a "mayor's project", to include the ambitions of their city into networks adapted to its means and needs.

This way of looking at metropolization has been accompanied in France by the increased prominence of a questionable but prevalent representation in the political debate: the divide between France's cities and peripheral France (Guilluy 2014). This schema has been criticized for its overly brief geographical determinism, in view of the social fractures that cut across the whole country and are moreover a response to socioeconomic criteria than to strictly territorial factors (Pech 2019). Another fracture line was proposed, replacing the opposition between cities and their peripheries with that of the integration (or not) of a territory through a metropolization dynamic. In other words, the updating of "productive-residential systems", where the economic dynamism of the city is based on the residential attractiveness of its countryside (Davezies and Talandier 2014). By this measure, 80% of the French population can be considered to live in an economically viable territory, even if it is exposed to future disruption. Therefore, the Bordeaux city is included in productive, non-market, yet dynamic France, with incomes that include tourism, pensions and public wages (Davezies 2012). Finally, Christophe Guilluy's opposition between city and periphery is challenged because it re-conduces, albeit in an apocalyptic rather than an apologetic register, the mythology that only cities are able to concentrate the highest employment and income when companies located in the periphery find their place in global competition (Bouba-Olga 2019).

In addition to the media benefits offered by its Manicheanism and its predisposition to unite anger, this discourse on the peripheral divide has been taken up by rural mayors to denounce the confiscation (in fact more rhetorical than effective) of public investment by cities. A heresy, according to them, with regard to the need to support the vitality of local exchanges, independent of all considerations for international competitiveness. "Open a retirement home, and you will save a school", rural elected officials can thus convince themselves (Darmian 2019). For those who remain convinced that, even if invisible, the dynamics of metropolization remain our goal (Offner 2018), it is in this context that a second figure of metropolization takes on allegorical value: the estuarine city. Derived from the Latin *aestuarium*, "place where the flow penetrates", the estuary is itself an allegory of metropolization. "A mixture of permanence and change," "massive and fragile at the same time", as Anne-Marie Cocula and Éric Audinet illustrate in their *Histoire de l'estuaire de la Gironde* (2018); the estuary can symbolize the connections of short- and long-term goals, of coastal shipping and long-distance shipping, of commercial exchanges, and agricultural or artisanal production. Now, this imperative to reconcile the characteristics and resources of the countryside with the appeal of the ocean had become the principle of the most vigorous criticism of "global cities", or more precisely of their elites, who were accused of neglecting relations with their countryside in order to focus on the connections between global cities alone. As if the perennial exploitation of agricultural and industrial activities no longer concerned them, as long as they could afford the products, and occasionally enjoy the charms of the countryside. Since the beginning of the 21st century, the metropolitan elite have rediscovered the importance of the territories that surround them, and seem to have become aware that they have at their disposal resources that are indispensable to their quality of life, if not to their survival. However, the flood zones around the estuary, the Medoc coastline, and the heat islands of the Bordeaux city, are likely to be painfully affected by global warming. This is a deliciously lucid view of Bordeaux, an English city from the 12th to the 15th century, more oriented towards the Atlantic than towards its countryside.

6.2. From the conquest of land to the recognition of territories

In order to understand the relations between Bordeaux and its countryside, historical hindsight makes it inappropriate to speak of territories and more appropriate to speak of domains appropriated or conquered by the Bordeaux bourgeoisie and nobles. The historian Anne-Marie Cocula reminds us of the domanial filiation between the villas of late antiquity, the rural lords of the Middle Ages and the modern châteaux devoted to the cultivation of wine (Cocula and Audinet 2018). Many of today's wine châteaux are the result of the transformation of these ancient medieval villages that multiplied at the end of the Hundred Years' War (Lavaud 2000). The same pattern persisted for nearly 2,000 years on the banks of the Garonne and the estuary. A "Bordeaux" oligarchy, different according to the period, composed[5], exploited, developed and sometimes expanded, thanks to the financing of the work required to conquer land from the sea, a rich terroir whose agricultural surpluses – mainly vineyards – allowed them to satisfy their own needs, as well as to develop a commercial activity. In addition to the enjoyment of a country residence, the estate was a source of prestige; an expectation reflected in the increasing attention paid to the architecture of the buildings. Holding an important political and economic role in the city and fruitful

---

5 For the modern and partly contemporary period, the high clergy, the "Nobles of the Gown", the rich cosmopolitan bourgeoisie made up of magistrates, guilds and traders who rubbed shoulders or snubbed each other, allied themselves (through marriage) or competed with each other, and occupied the Parliament, the Justice and the Royal Court.

possessions in the countryside, the accumulation of offices and having various sources of income, both commercial and agricultural, became the keystone for the success of the Bordeaux elite. This oligarchy, divided between town and country, has always shown a penchant for enjoying the calm of the countryside in the comfort of a town house, whether it be the *princes de la vigne* ("princes of the vineyard") or the *aristocrates du bouchon* ("aristocrats of the cork"). On the scale of history, more than its relationship of interdependence, Bordeaux maintains a relationship of monopolization, even predation, with the Gironde estuary. The estuary is in the service of the city of Bordeaux, to protect or feed it.

The shift from the categories of domain to those of territory marks the assertion of a preference for a republican rather than a despotic regime. The former aims at the general interest, the latter at that of the despot. The republic has been less imposed in so much as that despotism has been rejected. By evolving towards the confiscation of public power for private ends, the medieval lord ended up merging with the latter, with the peasants suffering under a less protective and increasingly spiteful lord.

In order for the obsessive formula of territorial equality to make sense, the category of territory had to cease to be a vague description of an expanse of land more or less characterized by geographical markers or the presence of a population, in order to become a space subject to a political institution whose borders limited the competence of the rulers. This was not yet the case in the Middle Ages, when feudal domination was based on the control of privileged places (grain storage silos, wells, mills, etc.), the existence of personal ties of dependence and manifestations of a symbolic order (Noizet 2016). As with most Western lords, the Bordeaux feudal lords of the 13th century were not all together; they were made up of a constellation of fields, villages and hamlets, sometimes very far from one another, but linked by a common dependence on a castle or a monastery whose name the lords bore; we could cross paths within the same town with men subject to different dominations (Boutruche 1947). The analogies are striking between this conception of space in the early Middle Ages and current invitations to better accompany the reticular dimension of metropolization: multipolar archipelagos energized by the circulation of people and goods, orchestrated on a regional scale by churches and castles, points of convergence of different networks; the difficulty of distinguishing the city from the countryside. In the Middle Ages, as at the end of the 20th century, "it is not space that defines a territory, but attachments, living conditions" (Latour

2018, *author's translation*). These two periods call for a combination of distance and mobility, which in the Middle Ages is illustrated by the ambiguity of the word *spatium*, "which could designate both duration, a lapse of time, and a spatial area, or the many terms designating surfaces by means of a temporal lexicon, such as '*journal*'[6] for the plot of land whose ploughing occupied a day" (Mazel 2016, p. 21, *author's translation*). Can we push the hypothesis that the estuarine city is part of a new movement of deterritorialization–reterritorialization comparable to what had been the transition from a mosaic of lords, controlled by a military aristocracy or by ecclesiastics, to the emergence of national territories?

A territory is not only what delimits the jurisdictional authority of an institution, it is also the object of its government. In France, this territorialization of space was largely initiated by the episcopal power. However, contrary to what historians taught in the 19th century, the ecclesiastical constitution of the dioceses was not a replica of ancient cities; the bishops did not first govern a territory with well-defined borders, but instead led a scattered group of faithfuls and churches through rituals and personal ties. Their power was challenged by the dominium of the great secular lords and the monastic communities that resisted their tutelage. It was through the Gregorian reform, which sought to affirm the role of the pope and to provide a better framework for secular society, that a process of territorialization by the dioceses took place, from the 11th to the 13th century. In the 14th century, the diocese was presented as an intangible space over which, apart from the monasteries, the bishop exercised dominion. This unprecedented sovereignty combined a tax system of its own with a legal system that was meant to apply to all the inhabitants of the territory, and not just a specific group. This territorialized relationship to the people would be taken up by princely and monarchical states to govern. If a religion can be defined as a successful sect, the territorial state is similar to a lord that has triumphed over its competitors. This vast territorialization movement by feudal lords contends with the identity myths based on the ancestral character of territorial boundaries (Mazel 2016).

This territorialization did not favor the condition of the most modest. This is what historian Florian Mazel observed in the Bordeaux region in the 11th and 12th centuries, when demographic pressures and the struggle between feudal lords pushed them to more regular use of the *saltus*, the more or less

---

6 French for "day".

wooded lands that the Gallo-Romans considered uncultivated and wild. "The control of traffic, the regulation of access to water or wood, the practice of fishing or the construction of mills made manifest the growing hold of the lords over spaces that had been shared amongst the peasantry up until then" (Mazel 2014, p. 478, *author's translation*). By creating territories, the lords extended their yoke when the *alleux*, the land without a master, diminished from the year 1,000. This same movement of territorialization led, in the middle of the 14th century, to the emergence of the city as a unitary place, distinct from the countryside, because it brought together within its walls what the Romans called the *urbs* (the built-up area of the city) and the *civitas* (its social community) (Noizet 2014). For the most vulnerable, the prejudices associated with the disorganization of the domanial regime seem to outweigh the benefits expected from the evolving organization of a territorial regime. By taking refuge in Bordeaux, a Bordeaux serf could gain their freedom if their lord did not claim them after a month and a day, however, at the price of abandoning their property to the domain because, unlike the Roman slave, the serf sometimes had a heritage that could be transferred to their children, a form of tenure for the land they worked (Boutruche 1947). In the end, the transfer by the nobility of part of its lands created day laborers who were more dependent on their new bourgeois owners than the hereditary tenants of the old domains:

> Therefore, in France, a people of workers without assured means of existence is created, living on a low and irregular salary. In the countryside they are day laborers who toil at the discretion of the owners, in the cities, workers work at the discretion of the masters (Seignobos 1982 [1933], p. 161, *author's translation*).

The Bordeaux region illustrates how the wealth produced by land clearing has increasingly benefited only a privileged few.

6.3. From equality to territorial cohesion

The French Revolution accomplished this abolition of privileges and instituted a first conception of territorial equality, which today appears to be outdated and yet continues to haunt us. This territorialization of space is accompanied by the idea that a society is all the more advanced the more the work is divided. By this measure, its growth becomes a necessary but not sufficient condition for its progress. A society can stagnate at a lower level of organization if its extension consists of an aggregation of watertight segments. The social horizon of individuals thus remains narrowly circumscribed and the division of social labor impeded. With regard to France, Émile Durkheim had no doubt that "the number and rapidity of the channels of communication and transmission" ensured the necessary fusion of small territories into the larger territory. He thus predicted, at the end of the 19th century, that the customs of the larger cities, less subjected to tradition and more open to progress, would spread to the smaller towns and then to the countryside, thus ensuring the hegemony of urban civilization (Durkheim 1978 [1898], p. 241, *author's translation*). In a Republic that conceives of itself as one and indivisible[7], the equality of territories is expressed first and foremost by an egalitarian institutional division of the national territory, as symbolized by the famous rule of being able to reach, on horseback, the prefecture of one's department in a day round trip. While fighting against the *terroirs* (Weber 1983), the republican state inherited from the French Revolution a network of its territory (*commune*, *department*, and then *region*), which had taken care to accommodate the communities of inhabitants (towns, parishes, etc.) of the Ancien Régime; this is the origin of the initial 44,000 *communes*. In this organization, which favored rurality in terms of political representation, Jacobinism always appeared tamed (Grémion 1976), and the "territorialization of politics" was understood as the adaptation and negotiation of central norms to local issues and contexts, particularly relationships to clients. France in the 20th century was not so much characterized by the presence of a strong state and by the centralization of decisions as by the interweaving of the central and the local, through which republican territories appeared to be in the hands of "barons". Therefore, Jacques Chaban-Delmas, deputy-mayor of Bordeaux, with a national political influence that made him Prime Minister and President of the National Assembly, could conveniently keep the same prefect for 14 years and decide on the appointment of directors of departmental administrations. The development of a territory depends mainly on the political will of these great chieftains, whose influence determines which projects to invest in (bridges, highways, industrial decentralization). This is not always of obvious rationality (the stopovers in a modest station conditioned by the notoriety of an elected official), nor crowned with success (the failure of the creation of a deep-water port at Verdon); however,

---

7 In reaction to the "unconstitutional grouping of disunited peoples" that was, according to Mirabeau, the France of the Ancien Régime.

in Aquitaine, Chaban-Delmas did not exercise his power exclusively for the benefit of Bordeaux; he was not averse to serving its periphery, the medium-sized cities and countries of the region. In the 1980s, when decentralization changed the rules of the game, these major elected officials seemed disconcerted by the heightened competition between territories (Oblet 2005), as well as by the obligation to take into account legislation that was less flexible to past arrangements.

Since the beginning of the 21st century, Bordeaux's elected officials have been working on a metropolitan project with the ambition or pretension of curbing the environmentally unsustainable and socially deleterious effects of anarchic urban sprawl. Its initial model is based on classic praise of density, valued for its supposed virtues in terms of energy sobriety, and an objective of demographic growth, presented at the time by Vincent Feltesse, president of the CUB, as an "opportunity" and not a nuisance, according to the principle: "More people mean more strength, more resources, more ideas, more opportunities as well" (Feltesse 2012, *author's translation*). The famous "Millionaire City" by 2030 is presented as a threshold status necessary to exist in the global world. The first effect of this project was to intensify competition between the territories, the city and the department, then the CUB and the General Council, to attract demographic and urban growth to them. In the second decade of the 21st century, this fracture between the department and the city is showing signs of healing. Within the framework of the InterSCoT[8] process, the departmental council recognizes the driving dimension of the city. The urban community recognizes that a large city must rely on competitive secondary centers and that an urban project in a rural area can curb rather than promote urban sprawl. The idea of metropolization, by dissociating itself from the sole obsession with urban density, raises awareness of the importance of dealing less with places than with links. Led by the Bordeaux urban planning agency, the construction of a sustainable mobility system is at the heart of discussions between the city and the surrounding territories (Godier et al. 2018).

In this context, the interdependencies reflected between the territories remain centered on the evolution of lifestyles and the dispersion of an individual's activities over several territories. The contrast between places of

---

8 Mission authorizing the meeting of elected officials and technicians of the department in order to coordinate the nine SCoT (*Schémas de Cohérence Territoriale*: Territorial coherence schemes) of Gironde, including that of the Bordeaux metropolitan area.

residence scattered over the entire department and jobs concentrated in the CUB forms the core of the debate. The categories mobilized to think about the interdependencies between territories are social in nature and concern facilities, housing, jobs and services. In short, we reason in the register of human resource management in a way that is quite similar to the way in which we became aware of the interdependence between people at the time of the industrialization of society.

The reflexive turn lies in discrediting the idea of territorial equality, understood as a sprinkling of public facilities and services, and the implementation of financial equalization mechanisms (more vertical than horizontal) intended to realize the old dream of a uniform France. This discrediting is very clear in *Révolution* (Macron 2017 [2016], p. 154). This conception of territorial equality no longer appears as the means to accomplish the economic and social equality of citizens, which it symbolized, in the "republican vision", when the population appeared less mobile and therefore easier to target in terms of investments. Far from the homogenization expected by territorial equality, the mobility of people pushes territories to compete with each other in order to attract the consumers of equipment and services that citizens have become. The defeated territories then deplore the fact that they have been forgotten by the Republic. However, if we reason in terms of waiting time, and not in terms of distance or the number of civil servants mobilized per inhabitant, rural areas receive far fewer services when compared to metropolitan areas. Equality and equity still stand out. As Philippe Estèbe develops, in addition to its very high cost, "territorial equality can lead to considerable social inequalities" (Estèbe 2015, p. 84, *author's translation*).

With regard to the redesign of solidarity policies, for 30 years sociologists have been warning about the need to target the public rather than the territories. The reflection emerged in relation to the suburbs, where the question was raised as to whether it was more judicious to break up ghettos than to help people get out of them. Inspired by the American example, Jacques Donzelot has suggested that focusing on the personal rights of the poorest would be more effective than aiming to transform the places where they reside (Donzelot et al. 2003). The inability of social rights (social citizenship) to guarantee the dignity of populations trapped in territories of relegation (in both rural and urban areas) led him to conceptualize, under the term urban citizenship – paraphrase here as metropolitan citizenship – the project of increasing equality of opportunity among individuals, which could extend the project (the project of social citizenship) of satisfying vital needs (Donzelot 2009). Criticizing the notion of territorial equality as a priority objective of policies carried out in the name of solidarity and social cohesion does not make us blind to territorial inequalities, understood as the fact that a place of residence is today less a marker of social dignity than an asset or a disadvantage in terms of access to education, employment, leisure and peace. Beyond the quarrels over the relevance of retaining this semantics, is it not this new conception of territorial equality that constituted the issue of the report directed by Éloi Laurent, which was supposed to inspire the action of the *ministère de l'Égalité des territoires et du Logement* (Ministry of Territorial Equality and Housing) (Laurent 2013)? Territories can hinder or promote the ability of individuals to become subjects of their own lives. The reference to Amartya Sen's notion of *capability* invites us to explore, in addition to income and GDP per capita, the non-monetary dimensions of human development (health, education, environment), the "real-life" indicators. Is it because of the overly strong shadow of the claims associated with the old conception of territorial equality that the expression seems to have lost its rhetorical force? On January 1, 2020, the *Commissariat général à l'égalité des territoires* (CGET: General Commission for Territorial Equality) was renamed *Agence nationale de la cohésion des territoires* (ANCT: National Agency for Territorial Cohesion). However, the expression remains a reference for rural mayors, anxious to defend their power against the various encroachments of inter-communality and to be able to count on an enhanced financial equalization.

This revision of the guiding principles of territorial equality permeates the most recent essays on the conditions of a new territorial pact (Viard 2018). This appears to be defined as a right to the city for all, a metropolitan citizenship of sorts. "The goal is not to have everything everywhere, but for everyone to have access to the same services and the same uses everywhere" (Viard 2019, p. 160, *author's translation*). For example, could "peripheral" communities not have residences built in the city that would allow their citizens to access certain metropolitan amenities such as the university, or develop a specific transport system for expectant mothers affected by the closure of a nearby, but at-risk, maternity hospital? Jean Viard seems to make the construction of this right the primary responsibility of regions (2018, p. 64). But can a region without a city be viable? Jean Viard's right to the city intends, more generally, to transform the bulk of regional policies "into horizontal policies of exchange and complementarity between the city and non-city for food, leisure, education, health, culture, energy, purifying the air and water" (Viard 2019, p. 159, *author's translation*). The possibilities associated with the fact that we no longer simply think of the relationship between the city and its surrounding territories in terms of human resource management, but combine this with the question of natural resource management, also emerge here. Nature is no longer simply a framework in which the history of human beings unfolds, but first and foremost the vital issue of their survival and, in a way, it is no longer secondary, but instead a source of wealth and the matrix of a reconfiguration of social relations. This analysis converges with Philippe Estèbe's valorization of a new territorial contract working towards new solidarities between territories that take into account that "tomorrow, low-density spaces will perhaps represent a decisive asset in a world where the sustainable management of natural resources will provide the basis for a new wealth of nations" (Estèbe 2015, p. 52, *author's translation*).

This observation and this outlook led us to move away from the alternative between the driving and predatory features of the city. The challenge is less to impose redistribution of resources between territories, but to see how these mechanisms can be combined in order to achieve a more equitable sharing of wealth. This implies paying a fairer price for the ecosystem services rendered by these territories, which could thus serve to strengthen the equal opportunity of the populations that have up until now been the most cut off from metropolitan opportunities, when interdependence and cooperation were viewed solely in terms of the social and economic perspectives. This re-composition is not without analogy with the relations that the West has maintained with developing countries, which could feel robbed in the exploitation of their natural resources.

The changes underway are still tentative and nebulous. The simple inclusion, in the key data of the *Métroscopie bordelaise* (A'urba 2019), of the very great dependence of Bordeaux Métropole for energy resources coming from outside its territory is, however, an indication of this. Not long ago, the worrying mention of this low level of independence (6.5%) would have made people smile, like a naive dream of self-sufficiency unsuited to large cities. Today, this dependence appears to be a price to be paid that needs to be better anticipated. To what extent can taking natural resources into account, in addition to human resources, broaden the scope of political negotiations between elected officials in the city and those in the countryside?

6.4. Conclusion

An allegory does not make a policy. More than delivering a new model of cooperation between territories, the Estuary City Model reveals all its complexity and potential for innovation. The challenge remains considerable. In the scientific literature devoted to the estuary, the alteration of the ecosystem by anthropic pressures maintains that the city is a problem (Sautour and Baron 2020). Between humans and wetlands, we would have to choose! The relationship between the estuary and its city appears to be full of ambiguities due to the constraints that we place on the other, and vice versa. In the Gironde, in addition to the asymmetry of the relationships of influence between the actors who are supposed to cooperate and the disparity of engineering resources, there is the persistent legacy of strong compartmentalization between the elected officials of the city and those of the countryside (Savary 2000). In view of the dreaded cumbersomeness of contractual formulas, which are supposed to initiate new relations of cooperation between urban and rural areas, but whose capacity to value ecosystem services has yet to be assessed, can the search for political devices better able to "represent flows and networks" and to "reorganize the technical–political work" (Offner 2020, *author's translation*) avoid institutional reform, despite the discredit to which the activity of institutional design is currently subjected? If a third symbolic issue were to be associated with the estuarine city, it might be to provide a breeding ground for inter-territoriality. Instructed in the pitfalls of the "territorial optimum", its mission would be to reduce the stiffness between, on the one hand, the protective solidarity expected, even passively, from a territory, where borders define the limits of an authority's legitimacy to exercise its power on the register of command and consenting obedience, and, on the other hand, the creativity of networks, where power is built on the register of negotiation. The estuarine city could initially, in the perspective of thinking of the national territory as composed of urban regions, correspond to the fusion of the *department* and the city, with subdivisions represented by a few elected officials from the rural and urban worlds, in a scheme evocative of the *demes* of the Greek cities of antiquity, which brought together citizens of the city and the countryside. The challenge would be to enable representatives of the rural world to be fully aware of urban issues, and vice versa. Like any territorial reform, it would come up against the corporatism of local elected officials and State or territorial civil servants, but nevertheless, an allegory must maintain reasons for hope.

6.5. References

A'urba (2019). Métroscopie bordelaise. Chiffres-clés [Online]. Available at: https://www.aurba.org/wp-content/uploads/2020/02/Metroscopie_bordelaise_2020.pdf.

Baczko, B. (1988). Lumières. In *Dictionnaire critique de la Révolution française*, Furet, F. and Ozouf, M. (eds). Flammarion, Paris.

Barber, B.R. (2015). *Et si les maires gouvernaient le monde ? Décadence des États, grandeur des villes*. Rue de l'échiquier, Paris.

Bouba-Olga, O. (2019). *Pour un nouveau récit territorial*. PUCA, Paris.

Boutruche, R. (1947). La crise d'une société : seigneurs et paysans du Bordelais pendant la guerre de Cent Ans. *Annales. Économies, sociétés, civilisations.* 2e année, 3, 336–348.

Cocula, A.-M. and Audinet, E. (2018). *L'estuaire de la Gironde, une histoire au long cours.* Confluences, Bordeaux.

Darmian, J.-M. (2019). *Le partage du pouvoir local*. Le bord de l'eau, Lormont.

Davezies, L. (2008). *La République et ses territoires. La circulation invisible des richesses*. Le Seuil, Paris.

Davezies, L. (2012). *La crise qui vient. La nouvelle fracture territoriale*. Le Seuil, Paris.

Davezies, L. and Talandier, M. (2014). *L'émergence de systèmes productivo-résidentiels. Territoires productifs – territoires résidentiels : quelles interactions ?* La Documentation française, Paris.

Desjardins, X. and Estèbe, P. (2019). *Villes petites et moyennes et aménagement territorial, éclairages anglais, allemands et italiens sur le cas français*. PUCA, Paris.

Donzelot, J. (1984). *L'invention du social, essai sur le déclin des passions politiques*. Fayard, Paris.

Donzelot, J. (2006). *Quand la ville se défait. Quelle politique face à la crise des banlieues ?* Le Seuil, Paris.

Donzelot, J. (2009). *Vers une citoyenneté urbaine ? La ville et l'égalité des chances.* Éditions rue d'Ulm, Paris.

Donzelot, J., Mével, C., Wyvekens, A. (2003). *Faire société. La politique de la ville aux États-Unis et en France*. Le Seuil, Paris.

Dubet, F. (2019). *Une vie de sociologue*. Le bord de l'eau, Lormont.

Durkheim, E. (1978). *De la division du travail social*. PUF, Paris.

Estèbe, P. (2015). *L'égalité des territoires, une passion française*. PUF, Paris.

Faburel, G. (2018). *Les métropoles barbares. Démondialiser la ville, désurbaniser la terre*. Le passager clandestin, Lyon.

Feltesse, V. (2012). *La Décennie bordelaise. Quelle politique urbaine à l'heure des Métropoles ?* Éditions de l'Aube, La Tour d'Aigues.

Godier, P., Oblet, T., Tapie, G. (eds) (2018). *L'éveil métropolitain, l'exemple de Bordeaux*. Le Moniteur, Paris.

de Godoy Leski, C. (2021). Vers une gouvernance anticipative des changements globaux. L'emprise des interdépendances socioécologiques sur une métropole estuarienne. Le cas de Bordeaux et l'estuaire de la Gironde. PhD Thesis, INRAE/Université de Bordeaux, Bordeaux.

de Godoy Leski, C., Gaillard, M., Sierra, M., Simonet, G., Bosboeuf, P. (2019). Regards interdisciplinaires pour une meilleure adaptation territoriale aux changements climatiques. *Natures, Sciences, sociétés*, 27(2), 212–218.

Grémion, P. (1976). *Le pouvoir périphérique. Bureaucrates et notables dans le système politique français*. Le Seuil, Paris.

Guilluy, C. (2014). *La France périphérique. Comment on a sacrifié les classes populaires*. Flammarion, Paris.

Latour, B. (2018). Il faut faire coïncider la notion de territoire avec celle de subsistance. *Le Monde*, 23 July.

Laurent, E. (ed.) (2013). *Vers l'égalité des territoires*. La Documentation française, Paris.

Lavaud, S. (2000). L'emprise foncière de Bordeaux sur sa campagne : l'exemple des bourdieux (XIVe–XVIe siècles). *Annales du Midi : revue archéologique, historique et philologique de la France méridionale*, 112(231), 315–329.

Macron, E. (2017). *Révolution*. XO Pocket, Paris.

Mazel, F. (2014). *Féodalités 888–1180*. Belin, Paris.

Mazel, F. (2016). *L'évêque et le territoire. L'invention médiévale de l'espace (V^e–XIIIe siècle)*. Le Seuil, Paris.

Noizet, H. (2014). La ville au Moyen Âge et à l'époque moderne : du lieu réticulaire au lieu territorial. *EspacesTemps.net*, 7 October.

Noizet, H. (2016). L'émergence de la ville au Moyen Âge. In *Quoi de neuf au Moyen Âge ?*, Catteddu, I. and Noizet, H. (eds). La Martinière, Paris.

Notre affaire à tous (2019). *Comment nous allons sauver le monde. Manifeste pour une justice climatique.* Massot, Paris.

Oblet, T. (2005). *Gouverner la ville.* PUF, Paris.

Offner, J.-M. (2018). *Métropoles invisibles. Les métropoles au défi de la métropolisation.* PUCA, Paris.

Offner, J.-M. (2020). *Anachronismes urbains.* Presses de Sciences Po, Paris.

Pech, T. (2019). La fin de la paix territoriale ? *Ihédate, terra nova*, 31 January.

Sassen, S. (1996). *La ville globale : New York, Londres, Tokyo.* Descartes et Cie, Paris.

Sassen, S. (2016). *Expulsions, brutalité et complexité dans l'économie globale.* Gallimard, Paris.

Sautour, B. and Baron, J. (eds) (2020). *L'Estuaire de la Gironde, un écosystème altéré ? Entre dynamique naturelle et pressions anthropiques.* PUB, Bordeaux.

Savary, G. (2000). *Chaban maire de Bordeaux. Anatomie d'une féodalité républicaine.* Aubéron, Bordeaux.

Seignobos, C. (1982). *Histoire sincère de la nation française.* PUF, Paris.

Stoléru, L. (1982). *La France à deux vitesses.* Flammarion, Paris.

Veltz, P. (2017). *La société hyper-insustrielle. Le nouveau capitalisme productif.* Le Seuil, Paris.

Vermeren, P. (2019). *La France qui déclasse. Les Gilets jaunes, une jacquerie au XXIe siècle.* Tallandier, Paris.

Viard, J. (2018). *Redessiner la France. Pour un nouveau pacte territorial.* Fondation Jean Jaurès, Paris.

Viard, J. (2019). *L'implosion démocratique.* Éditions de l'Aube, La Tour d'Aigues.

Vingré, M. (1980). *Le social c'est fini ! La nouvelle politique sociale : austérité, discipline, retour au marché.* Autrement, Paris.

Weber, E. (1983). *La Fin des terroirs. La modernisation de la France rurale 1870–1914.* Fayard, Paris.

7

Nantes and the Loire: Construction of an Estuarine City Faced with Port and Environment Challenges

While French cities can be coastal (Marseille, Brest, etc.) or mountainous (Grenoble, etc.), they can also be estuarine (Nantes, Bordeaux, Rouen, etc.). These qualifiers call for the inclusion of cities within an estuarine biophysical environment that goes beyond political–administrative perimeters. More precisely, under this notion of estuarine city, it is important to consider the question of the relations that these cities at bottom of the estuary maintain with the river and its river mouth, its tides and tidal bed, its wetlands, its secondary tributaries and its alluvial plain. On the one hand, this notion shows the importance of hydrographic and geomorphological conditions as "natural" elements structuring the urban fabric and development. On the other hand, it commits us to taking into account the consequences of the strong constraints exerted by the metropolization phenomenon on these environments. The objective is therefore to reflect, through the notion of the estuarine city, on the socio-ecological interdependencies between a city and its surrounding environment, by understanding the term environment as both a biophysical entity (ecology) and a socio-political space (territory).

Arguments developed in this chapter are based on the case of the Nantes metropolitan area and the Loire Estuary. Nantes, like Bordeaux and Rouen, is one of the three French cities at the bottom of the estuary located on the

---

Chapter written by Glenn MAINGUY.

Atlantic coast. It is located 60 kilometers from the mouth of the river. The Loire estuary has always been a place of intense human activities (fishing, agriculture, extractive industry, etc.). Nevertheless, this anthropization and profound transformation of its watercourse began in the 19th century. From 1850 onwards, intensification of estuarine development was justified by the creation of the first harbor basins in Saint-Nazaire and also by the silting up of the Loire bed. Works must facilitate navigation on the estuary for ships with increasingly large tonnages, and thus allow the city of Nantes to maintain its port vocation opposite Saint-Nazaire. In addition to the historical development of the Nantes–Saint-Nazaire industrial-port complex, there are now the effects of urban sprawl induced by the phenomenon of metropolitization. Demographic evolution of the territory bears witness to this strong economic attractiveness. Between 2012 and 2017, the city of Nantes gained 43,599 inhabitants, and projections for 2030 predict a population growth of 18% (AURAN – *Agence d'urbanisme de la région nantaise*: Urban planning agency of the Nantes region). The strong anthropic pressures on the Loire territory have resulted in a significant deterioration of the hydrological functioning of the estuary: "Significant water and air pollution, extension of the muddy plug, lowering of the river level at low water, increase in the salinity limit, loss of biodiversity" (Despres 2009, p. 15, *author's translation*). In the space of a century, between Nantes and Saint-Nazaire, the surface area of intertidal zones (tidal zones) has been halved. While the anthropization of the estuary territory intensified from the 20th century onwards, it was during the 1990s that the true extent of the ecological degradation was described, leading to the creation of the GIP (*Groupement d'intérêt public*) Loire Estuaire in 1998:

> The water lines are getting lower. In 1989, the drinking water intake of the Nantes agglomeration was moved because the salt was rising in the estuary. The farmers also observed changes in the submersion phenomena, the phenomena of water salinity, in the change of the particular flora used for grazing; also in the silting up of the channels, and thus the more important cleaning processes. The fishermen also observed fewer fish, and some even told us that when they put their hands in the water, they came out red (GIP Loire Estuaire Interview 2019, *author's translation*).

Today, due to heavy anthropization, the estuarine space of the Loire extends over a hundred kilometers, from the line formed by the tip of

Saint-Gildas and the tip of Chémoulin to the limits of the tidal wave point at Ancenis. This territory, which is called the "dynamic estuary", goes far beyond the boundaries of the 24 communes of Nantes Métropole to include all of the intercommunal areas surrounding the estuary: Nantes Métropole, agglomeration community of the Nazairian region and estuary, Estuaire and Sillon community of communes, Pays d'Ancenis community of communes, Mauges community of communes, Sèvre and Loire community of Sud Estuaire communes. This ecological definition of the upstream and downstream limits of the dike estuary is not well known by the stakeholders of the estuary. For many, the estuary stops at Nantes: "We tend to forget this part of the estuary upstream" (Voies navigables de France interview 2019, *author's translation*). For others, the upstream and downstream represent two disjointed environments that do not form a single biophysical entity:

> It is difficult with actors who systematically divide this estuary upstream/downstream from Nantes. Even in meetings, we are always forced to say, "What does the estuary mean to you?" People always ask us, "Are you talking about downstream? Upstream? Of the whole? Of what?'" (SAGE Estuaire de la Loire Interview 2019, *author's translation*).

This chapter analyzes the evolution of socio-ecological interdependencies between the city of Nantes and its biophysical estuarine environment of the Loire, and actors who participate in making them visible. The aim is to see how the estuarine dimension of the city of Nantes and the Nantes territory is embodied in the discourse and practices of institutional actors. Which narratives are carried and promoted by actors? Which narratives described in this chapter are not the only ones that exist; they are part of parallel histories and recurrent ways of constructing the link with the Loire estuary.

7.1. Emergence of the estuarine dimension: from the city of Nantes–Saint-Nazaire to the opening of the "Terre d'Estuaire" museum

7.1.1. *From an institutional dimension ...*

In the case of the Loire estuary, the term estuarine city appears to describe a territory that brings together two urban centers: Nantes and Saint-Nazaire. The emergence of the estuarine dimension of a metropolitan territory was not so much linked to environmental issues as to political and economic issues in the second half of the 20th century. This economic planning was initiated by the State within the framework of the creation of regional metropolises:

> In the 1960s, the State wanted to turn this coherent geographical ensemble into a balanced city. The watchword was simple: to balance Paris and its province by structuring an atrophied and isolated regional city; more broadly, to design the modern city of the time: an industrial city on the banks of the estuary. The State thus initiated the first meeting of Nantes and Saint-Nazaire between land and sea [...] The cooperation [...] around industry and port must make "war" impossible [...] and peace possible (Mahe and Bois 2012, p. 20, *author's translation*).

Therefore, the estuarine city was initially built around the creation of a major industrial-port pole following the merger of the port of Saint-Nazaire with that of Nantes in 1966, called the "autonomous port of Nantes–Saint-Nazaire". The energy vocation of the site, which was embodied in the presence of a refinery and a coal-fired power station, was accentuated by the installation of a methane terminal (Leroy 1996). During this period, the ports of Nantes and Saint-Nazaire played a dominant role in the constitution and management of the estuary territory. From the 1980s onwards, local politicians took matters into their own hands with the aim of institutionally embodying the estuarine dimension of the metropolitan pole. This embodiment was first a symbolic act. In 1989, Jean-Marc Ayrault, the newly elected mayor of Nantes, and Joël Batteux, mayor of Saint-Nazaire, sailed down the Loire estuary from Nantes to Saint-Nazaire together with their respective teams. In parallel with the rapprochement between the mayors of Nantes and Saint-Nazaire, the *Association communautaire de l'estuaire de la Loire* (ACEL) – bringing together the main economic and political decision-makers of the estuary (mayors, presidents of the region, General Council, Chambers of Commerce and Industry and Autonomous Ports) – was created in 1996. In 1999, the first metropolitan conference took place, bringing together various political (82 mayors present) and economic actors of the estuary. In 2003, a mixed syndicate was set up to carry out a territorial coherence plan (SCoT) of the metropolitan area. This SCoT was adopted in 2007 and is based on four pillars: combating urban sprawl, developing existing urban centers, combining mobility and the urban fabric, and safeguarding agricultural and natural areas (SCoT 2007). It includes 57 municipalities and five inter-municipalities, mainly located north of the estuary. A first limit to the integration of the estuary by the metropolitan pole project is observed by the absence of the territories south of the estuary, which have their own SCoT, and which is now the Pays de Retz territorial and rural pole and includes four communities of communes.

7.1.2. ... to a cultural and tourist vocation

During the 2000s, in parallel with development of SCoT, culture, heritage and art were mobilized in the creation of the estuary metropolitan territory (Coëffé and Morice 2013). Jean Blaise, then director of the *Lieu Unique à Nantes*, set up the first edition of the *la biennale d'art contemporain Estuaire* (Estuary contemporary art biennial) in 2007, which was extended in 2009 and 2012:

> From these three creative moments emerged a network of *in situ* works, scattered between Nantes and Saint-Nazaire, on the 60 kilometers on either side of two banks. While some creations were conceived to shape the landscape during the ephemeral time of the event, others were conceived according to a "sustainable" orientation, by deploying on the scale of the estuary a perennial course accessible to the public [...] The principle is to bring into existence, through art and its mediatization, a future metropolitan area of the Loire with a European influence of nearly one million inhabitants (Coëffé and Morice 2013, p. 81, *author's translation*).

In a complementary manner, the estuarine dimension of the metropolitan pole will be embodied in the development of tourist activities. We can mention the creation in 1996 of the *Estuarium* association, which presents itself as a lever for development of natural and cultural heritage. Its ambition is to participate in the construction of a common and shared identity, both north and south of the estuary. The incarnation of the estuarine dimension is achieved through highlighting the estuarine heritage. In 2011, a local public company, named *Le voyage à Nantes*, was created with the objective of providing "the estuary with an organization capable of promoting the territory to become one of the top tourist spaces in Europe" (Coëffé and Morice 2013, p. 84, *author's translation*). Extending this idea, in 2019, a museum of the territory called *Terre d'estuaire – La Loire de tous les sens* was opened in Cordemais. Below is the presentation proposed by the instigators of the museum:

> A pirate's safe haven, a starting point for voyages around the world, an idyllic setting for biodiversity, key to a rich industrial activity, the Loire estuary is a territory with a unique natural heritage, a small strip of 60 km that has so many stories to tell! Play the hunter of wrecks lying underwater, imagine yourself a migratory bird by lying on a sofa, improvise as a boatman or fly above the estuary in a balloon: *le Centre de découverte de Terre d'estuaire* offers the possibility to contemplate, discover, understand, experiment and travel this last part of the Loire, before it meets the ocean.

Through the development of its cultural and tourist vocations, the estuary is embodied. It "becomes a space that can be visited […] a woven fabric of places" (Coëffé and Morice 2013, p. 85, *author's translation*).

The development of the Nantes–Saint-Nazaire metropolitan cluster project expresses an initial sense of the link between urban development and territorial registration within the Loire area. In this integrative conception, the estuary is thought of as a backbone geographically linking two urban spaces with the objective of forming an estuarine metropolitan territory. If the dynamic is launched, achievements are still sketchy. Estuarine territorial identity is far from being adopted by inhabitants and political leaders – "Yes, rurality is more important than the estuary" (mayor of a municipality located south of the Loire estuary, April 2019, *author's translation*); "We feel, today, that the metropolis has not shown a genuine interest in working with us; it is more by obligation than by need" (Bahurel 2012, p. 158, *author's translation*) – or the economic actors of the estuary. In parallel to these difficulties, the city of Nantes, and consequently Nantes Métropole, sought to rebuild its relationship with the Loire at the start of the 1990s.

7.2. When Nantes and the Loire drifted apart: a progressive denial of the city's estuarine dimension

7.2.1. *From digging a canal to filling in waterways*

From the 1990s onwards, the city of Nantes, through its mayor Jean-Marc Ayrault, embarked on a process of reconciliation with the Loire, symbolized in particular by the "Île de Nantes" project. To understand importance of this resolution, it is useful to look back at the history of Nantes' relationship with the Loire during the 20th century. The 20th century marked a shift in the relationship between Nantes and its estuary. This distancing was caused by various estuary developments in the first part of the 20th century. At the end of 19th century, the port of Nantes was faced with a major problem of silting up the estuary. In the long term, this situation could severely limit access to the Nantes quays for ships with an ever-increasing tonnage and draught. At that time, two solutions were considered. Those in favor of the first solution proposed building a canal alongside the Loire. Those in favor of the second solution wanted to continue the work of dredging the Loire bed. When it became clear that the limits of the technical capacities of dredging were being reached, the first option was definitively adopted on March 3, 1882. Construction of the canal de la Martinière began, and the first boat used it 10 years later, in September 1892. Its use lasted 21 years. Faced with the increase in the size of merchant ships and the progress of dredging technologies (steam dredgers), the canal quickly became obsolete. In 1903, the option of developing the riverbed came back into force and dredging work in the estuary and port resumed (Maquet 1974, pp. 80–83). In 1913, the last ship used the Martinière canal. The same year, digging of the bed of the Loire had a new objective. They had to fight against the silting up of the port by creating a tidal basin upstream from Nantes. The purpose of the works was to increase the speed of water circulation during the ebb tide, in order to produce a self-draining mechanism that would carry sediments towards the ocean more efficiently (Maquet 1974, p. 80). The most visible consequence is the lowering of riverbed at Nantes by about 4 meters. "The tidal range is now greater in Nantes than in Saint-Nazaire" (GIP Loire 2019 interview, *author's translation*). Although these effects were foreseen by the civil engineers, others were less expected:

> Engineers […] had initially overlooked the fact that Nantes is, like Venice, a city built on wooden columns, due to the unstable nature of its sedimentary soil. As the water level dropped, foundation columns, which were intended to remain submerged, were exposed to air twice a day at low tide. They began to rot. Engineers had also overlooked erosion that would quickly occur in underpinnings of the wharves and slips due to the new strength of the current (Massard Guilbaud 2015, p. 72, *author's translation*).

Acceleration of the current and lowering of the water level at low water in Nantes put a lot of stress on the river buildings (quays, bridges, docks, etc.) and led to the collapse of several of them: Ernest-Renaud quay (1911), Maudit bridge (1913), Magellan quay (1924), Pirmil bridge (1924) and Petite-Hollande square (1925) and the collapse of the Turenne quay and Gloriette Island (1927–1928):

> In fact, according to [civil engineers], it was the entirety of the downtown wharves and bridges that were threatened with collapse in the near future (Massard Guilbaud 2015, p. 74, *author's translation*).

In order to quickly address this situation, national civil engineers directed Nantes municipality to fill both arms of the Loire and Erdre. Work began in 1926 and lasted 20 years. An important morphological transformation of the city took place. "The image of the Venice of the West was completely erased" (AURAN interview 2019, *author's translation*). Justification given at the time for accepting these large-scale works was quite different. Geneviève Massard Guilbaud (2015), looking back at history of the Nantes floods, shows how the natural risk of flooding was used a posteriori to legitimize work and decisions undertaken by civil engineers. The Loire and its estuary were presented as obstacles to the modernization of the city, and in particular to the development of new forms of mobility, such as trains and later automobiles. They were perceived as sources of squalor. And above all, omnipresence of water makes the city highly vulnerable to flooding. Three major floods hit Nantes in 1904, 1910 and 1919. Filling in of the river was justified as a solution to flooding, urban squalor and for opening the city, and not as a solution to the economic influence of the port and the functionality of the estuary itself, to guarantee the navigation of boats.

Today, stakeholders evoke a "history of departures", a river "that is going to be further away, deeper" and a loss of a certain number of uses linked to the proximity of the Loire:

> The Erdre bed is transformed into a major road artery (*le cours des B50-Otages*). The Petite-Hollande Esplanade (mainly a parking lot) is installed on an old arm of the Loire. Feydeau Island ceases to be an island. The Loire that used to pass in front of the castle no longer passes in front of the castle (Nantes Métropole interview 2019, *author's translation*).

7.2.2. Departure from shipyards

Forty years later, a second event affected Nantes' relationship with its watercourse. On July 13, 1987, the *Bougainville*, the last ship built in Nantes, cast off from Dubigeon shipyard, marking its closure. Dubigeon shipyards had been located in the heart of city since 1842, on the island of Prairie-au-Duc and in Chantenay-sur-Loire. In a little over a century, 1,200 ships were built there (ocean liners, sailing ships including the Belem, submarines, dredgers, cargo ships, trawlers, etc.). Shipyards reached their peak in the mid-1950s with more than 7,000 employees working on the site. The closure of the shipyard in Nantes was part of a larger movement, characterized by the relocation of naval and port industries from Nantes to Saint-Nazaire. With the closure of the shipyards, "the island of Sainte-Anne became a huge industrial wasteland and quays of the Madeleine arm were no longer beset by trade ships. This marked the beginning of a form of 'mourning' for the Loire by people of Nantes, who turned away from a 'faded' and lifeless river" (Grand Débat "*Nantes, la Loire et nous*" 2015, p. 41, *author's translation*).

7.3. Building a new relationship between Nantes Métropole and its estuary: the desire for the Loire

7.3.1. *Integration of the estuarine dimension through heritage and industrial-port memory*

Since the 1980s, the city of Nantes has been confronted with a problem common to river-port cities such as London, Bilbao and Bordeaux (Chaline 1992, 1993), that of the reconversion of spaces, often industrial wastelands, left vacant by the disappearance or relocation of important port activities linked to the maritime economy. How can we mobilize the heritage of an industrial past, "glorious" but weighty, in the face of current challenges? Two different visions of the urban project and of city–estuary interactions gradually emerge. The first vision was proposed by the municipal team of Michel Chauty. Following urban planning principles of the time, a project for an international business city was envisaged. A Luxembourg developer was found. Within this first urban project, no place was given to the Loire. The industrial and river world had to be erased to make way for new uses oriented towards the tertiary and globalized economy (Ghorra-Gobin and Reghezza-Zitt 2016). Dismantling of the site by scrap metal workers began in 1988 but was quickly halted following the intervention of former shipyard workers, who mobilized to protect their technical heritage (know-how, archives, buildings), and also to "safeguard the working-class and naval memory" of Nantes (Peyon 2000, p. 117, *author's translation*). The Loire and its estuary are thus thought of, above all, through the set of former industrial activities. More than a biophysical entity to be preserved from ecological degradation, they are considered elements of a heritage to be preserved and as memories of the territories. As Chasseriau and Peyon remind us, mobilization of former Naval shipyards puts "at the heart of urban reflection the will to preserve the riverine, maritime and industrial identity of this vast district" (2004, p. 45, *author's translation*).

Mobilization of former naval personnel met with a favorable response from the people of Nantes and was gradually coupled with a political will to bring the city closer to the Loire. "A diagnosis had shown [at that time] the importance of water for the people of Nantes" (Chasseriau and Peyon 2004, p. 45, *author's translation*). These two imperatives progressively impregnated the philosophy of the urban project, whose objective was to reestablish a link between the city and its estuary. The arrival of Jean-Marc Ayrault at the head of city council and the integration of former naval personnel into the municipal team in 1989, brought with it the desire to build a new relationship between the estuary and the city of Nantes. This desire is reflected in words of policies that affect territorial representation. From the 1990s onwards, the aim was to "reweave the link between the Loire and Nantes", to open up "access to the Loire" and thus to restore proximity to the river. This change of perspective to the Loire estuary is embodied in the "Île de Nantes" project, initiated by architect Dominique Perrault and developed from the early 2000s by Alexandre Chemetoff's architectural firm. As the project's presentation documents show, the latter aims to transform the Loire into the new urban centrality of Nantes:

> On the island, any development action refers directly or indirectly to the theme of the city's opening onto the Loire. It is on the island of Nantes where the relationship between the city and the river can be cultivated and give any development its meaning. Any action undertaken must satisfy the idea that it introduces, develops, and restores a relationship between the city and the water (Chemetoff and Berthomieu 1999, Preface, *author's translation*).

The urban project defended by Chemetoff's team, through the concept of a "recovered Loire", aims to articulate the memorial and patriotic imperatives defended by former members of the navy and the desire of citizens for more nature and more water:

> It is a question of finding a way, a living way to cultivate the memory of past activities that have marked the history of the relationship between the river and the city, and at the same time to ensure that the entire agglomeration develops in its geographical center the landscape of a city open to the river (Chemetoff and Berthomieu 1999, *author's translation*).

7.3.2. The *Great Debate: a participatory tool for reclaiming the Loire*

Urban rehabilitation work undertaken by Jean-Marc Ayrault's municipal team has made it possible to "see the Loire in a positive light again and the first developments of public spaces are reestablishing a closeness with the Loire" (Nantes Métropole interview 2019, *author's translation*). Johanna Rolland's successor as mayor of Nantes and head of the city in 2014 paved the way to a new stage in the construction of Nantes' relationship with the Loire. This relationship with the Loire involves a new question: What is the place of a river in a modern city? The new president of Nantes Métropole speaks of a "new ambition for the Loire" and "a reconquest of the Loire for all" (Rolland 2015, p. 3, *author's translation*). A reading of various reports related to this estuarine ambition and interviews conducted shows that this reconquest of the Loire is broken down into several objectives: a pacification objective, namely, "reconciling Nantes and the Loire" (Nantes Métropole interview 2019), *author's translation)*; a cohesion objective found in the desire to "reweave the link between the Loire and Nantes"; a mobility objective, namely, "making the Loire more accessible" (AURAN interview 2019, *author's translation*); a more subjective objective of appropriation, in other words "the wish, the desire, the dream of the Loire" (Loire 2020 Permanent Commission report, p. 8, *author's translation*). The story of this new relationship between Nantes and its river will be constructed through the organization of a Great Citizen Debate.

The Metropolitan Council deliberated in June 2014 in favor of organizing a Great Citizen Debate, named "*Nantes, la Loire et nous*". It began in October and lasted eight months. Its objective was to "involve residents in a reflection on the place and role of the Loire in the development of the territory" (*Nos engagements pour demain*[1] 2015, p. 4, *author's translation*). Nearly 5,000 people participated actively and physically, and 40,000 participated through the development of dedicated digital tools. The Great Debate was organized around four themes:

- the Loire of practices and uses;

- the Loire as an economic and ecological space;

- the Loire, mobility and crossings;

- the Loire, heart of the city, attractiveness and urban quality.

At the end of the Great Debate, elected representatives of the city made 30 commitments outlining "a new ambition for the Loire" (*Nos engagements pour demain* 2015, p. 12, *author's translation*). This ambition is embodied in 12 major projects that respond to six major themes identified by the participants in the Great Debate to determine "the place of a river in a 21st century city" (the Great Debate *Nantes, la Loire et nous* 2015, p. 41, *author's translation*). In the context of these major projects, leisure and culture are being mobilized with the aim of making the estuary their own. Several festive and recreational initiatives have been set up (*Fête nautique Débord de Loire*, guinguettes, etc.). Elected officials of the metropolitan area are committed to structuring footpaths (soft mobility) on the banks of the Loire and to building lookouts "that allow one to read the landscape, to read nature" (Nantes Métropole interview 2019, *author's translation*). The second identifiable issue is that of sustainable urban development and the search for a balance between economic growth and environmental preservation. On the one hand, the estuary is considered natural heritage in which economic activities must be maintained and developed (development of river transport, mobilization of port infrastructures for urban logistics needs), while seeking to limit effects and constraints exerted by these activities on the environment (zero phytosanitary action, biodiversity inventory, etc.). On the other hand, the Loire and its estuary are placed at the center of urban planning ambitions and major development projects in the city (Place de la Petite-Hollande, Bas-Chantenay, Malakoff), and are seen as major landmarks of urban attractiveness (the Loire as a new metropolitan center). Finally, a third identifiable challenge is that of mobility. The objective, through the profound transformation (enlargement and widening)

---

1 Our commitments for tomorrow.

of the Anne-de-Bretagne bridge, is to transform the estuary from a border into a link between the northern and southern banks of the Loire.

7.3.3. Conférence Permanente Loire *and* Mission Loire*: putting environmental issues related to the Loire on the agenda*

In addition to the 30 commitments made by the metropolitan elected officials, the Great Debate also made it possible to set up a monitoring and evaluation body in 2016: the *Conférence Permanente Loire*. The latter has three main missions. It monitors, gives its opinion on the implementation of the 30 commitments by the metropolitan elected officials and produces recommendations. These three objectives are materialized in the drafting of an annual report (January 2017, February 2018 and January 2020) evaluating the progress of the achievements. It is presented as a body for citizen participation and uses two participatory tools in particular: organization of citizen workshops and the holding of plenary public meetings. The *Conférence Permanente Loire* is made up of people from different sectors (metropolitan elected officials, citizens, experts, private players and associations). It had 19 members at its creation. Members of the *Conférence Permanente Loire* serve for four years. The first mandate ended in 2020. In 2021, a new team of 26 members was formed. By comparing the composition of two conferences, it is possible to make several conclusions. The team has grown. It has gained seven members. This increase is mainly due to the increase in the number of citizens (from three to eleven). The number of metropolitan elected officials has decreased. Beyond these two changes, the most striking inflection is in the general composition of the two conferences. The first conference was marked by the strong presence of people from the world of urban planning, development and transport. In the second conference, we can observe the integration of representatives of environmental issues: the *Ligue de protection des oiseaux de Loire-Atlantique* (Loire-Atlantique Bird Protection League), the Agence de l'eau (Water Agency) and the elected official in charge of the Loire – an illustration of a form of "ecologization" (Ginelli 2017) of metropolitan public action in favor of the Loire:

> We feel very clearly, through this *Conférence Permanente Loire*, a rise in these environmental concerns related to water quality, climate change, pollution issues (Nantes Métropole interview 2019, *author's translation*).

In addition to *Conférence Permanente Loire*, *Mission Loire* was created following the Great Debate. It has been set up with two objectives in mind. The first is identical to the mandate given to *Conférence Permanente Loire*, that of monitoring and evaluating the implementation of 30 commitments. Beyond this objective, *Mission Loire* was conceived as a response to the truncated representation of the Loire within metropolitan services. As the head of *Mission Loire* recalls, "it was in 2015–2016 that mission was set up […] because we saw that in administrative departments, the Loire was always represented by a small party. There is no continuity. It is always something vague. It is abstract. It's present, but it's a presence that's a bit diluted" (Nantes Métropole interview 2019, *author's translation*). The cause of the fragmented vision of the Loire lies in the scattered handling of various issues related to the Loire through a plurality of services.

Beyond the fragmentation of estuarine issues, stakeholders point to obstacles linked mainly to their professional culture and difficulties they have in appropriating the Loire and working with the estuary. The estuary is still a strange territory that is difficult to understand:

> It's not the usual playground for all services. This is true in our departments and it is also true among our planners, who are rather terrestrial. They tend to think: "We do everything right with the Loire. We'll go to the edge, but not too close" (Nantes Métropole interview 2019, *author's translation*).

A second objective of *Mission Loire* is to appear as a "coordinator", by acting as a link between various departments dealing with issues concerning the Loire and its estuary. Under the *aegis* of the General Directorate for Territorial Coherence, and thanks to the creation of a Loire technical team, *Mission Loire* carries out internal coordination work. In order to do this, it has two mechanisms for action: one is informative (centralization and circulation of information between various departments) and the other is consultative (drafting opinions on projects). Through the drafting of framework notes, mission letters and the production of a river planning document, the objective of *Mission Loire* is therefore to change the way in which metropolitan public action is viewed, taking the estuary as its point of entry: "the idea is to make it easier to take account of the river" (Nantes Métropole interview 2019, *author's translation*). It seeks to transform the way administrators and planners look at the Loire, by developing a professional culture based on the environment and river development: "We see the Loire from afar, and rather like landlubbers. We have to see it from the Loire" (Nantes Métropole interview 2019, *author's translation*).

7.4. Conclusion

By taking into account different territorial scales (Nantes Métropole and Nantes–Saint-Nazaire Ecométropole), this chapter has examined the ways in which institutional stakeholders participate in the visibility of socio-ecological interdependencies between an estuary and a city, in the context of global change. The socio-historical approach mobilized to study the narrative of these interdependencies shows the evolution of references for development of the Loire estuarine territory that were adopted to deal with the new ecological situation (Salles 2006). This movement is threefold. On the one hand, it is characterized by the entry of the estuary into politics (Latour 1999) via processes for institutionalization of estuarine issues at the level of the cities of Nantes and Saint-Nazaire. On the other hand, it is defined by an environmental inflection induced by a change of doctrine and a vision of what the Loire and its estuary are: no longer just a maritime route to bring boats up to Nantes, but now an ecological territory to be preserved and enhanced. "It is no longer a question of adapting the Loire to navigation, but of adapting to what the Loire is, and to its extremely important ecological issues" (VNF interview 2019, *author's translation*). Finally, this environmental inflection is accompanied by the desire to give a new status to the Loire and its estuary in the making of the city, by conceiving them as new urban centralities. These results can be compared with observations made in other estuary territories (Gironde, Seine), and thus serve to compare trajectories of estuary cities in the context of global change. The trajectory of the Seine estuary remains, as shown in the work of Gilles Billen and his collaborators (see Chapter 8), determined by port interests and the maritime transport of goods. The estuarine territory here is completely interwoven with the larger "Seine axis", characterized by the merger of the ports of Le Havre, Rouen and Paris (Haropa). Bordeaux, as a flood-prone city, conceives of its geographical position as a vulnerability (de Godoy Leski 2021). Its estuarine dimension is mainly built around taking into account and combating flood risks, to the detriment of a global consideration for environmental concerns within the territory.

7.5. References

Bahurel, M. (2012). Paimboeuf : pour une métropole vraiment estuarienne. In *Estuaire Nantes–Saint-Nazaire. Écométropole mode d'emploi*, Masboungi, A. (ed). Le Moniteur, Paris.

Chaline, C. (1992). Le réaménagement des espaces portuaires délaissés : une nouvelle donne pour la centralité urbaine. *Les annales de la recherche urbaine*, 55/59, 79–87.

Chaline, C. (1993). Réflexion sur la reconquête des waterfronts en Grande-Bretagne. *Norois*, 40(160), 589–599.

Chasseriau, A. and Peyon, J-P. (2004). Le projet île de Nantes, ou comment la ville se réconcilie avec son fleuve. *ESO Travaux et Documents*, 22, 41–50.

Chemetoff, A. and Berthomieu, J.-L. (1999). *L'Île de Nantes, le plan guide en projet*. Memo, Nantes.

Coëffé, V. and Morice, J.-R. (2013). Patrimoine et création dans la fabrique territoriale : l'estuaire ligérien ou la construction d'un territoire métropolitain. *Norois*, 3(228), 77–88.

Despres, L. (ed.) (2009). *L'estuaire de la Loire. Un territoire en développement durable*. Presses Universitaires de Rennes, Rennes.

Ghorra-Gobin, C. and Reghezza-Zitt, M. (2016). *Entre local et global : les territoires dans la mondialisation*. Le Manuscrit, Paris.

Ginelli, L. (2017). *Jeux de nature, natures en jeu. Des loisirs aux prises avec l'écologisation des sociétés*. Peter Lang, Brussels.

de Godoy Leski, C. (2021). Vers une gouvernance anticipative des changements globaux. L'emprise des interdépendances socioécologiques sur une métropole estuarienne. Bordeaux Métropole et l'estuaire de la Gironde. PhD Thesis, Université de Bordeaux, Bordeaux.

Latour, B. (1999). *Politiques de la nature. Comment faire entrer les sciences en démocratie*. La Découverte, Paris.

Leroy, M. (1996). Comment concilier économie et écologie dans l'estuaire de la Loire. *La Houille Blanche*, 6/7, 89–91.

Mahe, S. and Bois, S. (2012). Naissance d'un grand territoire. In *Estuaire Nantes–Saint-Nazaire. Écométropole mode d'emploi*, Masboungi, A. (ed). Le Moniteur, Paris.

Maquet, J.F. (1974). L'aménagement de l'estuaire de la Loire. *La Houille Blanche*, 1/2, 79–89.

Massard Guilbaud, G. (2015). Du risque naturel comme outil de légitimation de l'aménagement territorial. In *Les Territoires du risque*, Granet Abisset, A.-M. and Le Gal, S. (eds). Presses universitaires de Grenoble, Grenoble.

Nantes Métropole (2015a). Nos engagements pour demain. Report, Nantes Métropole, Nantes.

Nantes Métropole (2015b). Le Grand Débat. Nantes, la Loire et nous. Report, Nantes Métropole, Nantes.

Nantes Métropole (2017). Conférence Permanente Loire. Report, Nantes Métropole, Nantes, January.

Nantes Métropole (2018). Conférence Permanente Loire. Report, Nantes Métropole, Nantes, February.

Nantes Métropole (2020). Conférence Permanente Loire. Report, Nantes Métropole, Nantes, January.

Peyon, J.-P. (2000). Patrimoine et aménagement urbain à Nantes : des relations conflictuelles permanentes. *Norois*, 87(185), 113–123.

Pinson, G. (2009). *Gouverner la ville par projet*. Presses Universitaire de Sciences Po, Paris.

Renard, J. (2013). Nantes 2030, observations sur une réflexion prospective citoyenne [Online]. Available at: https://cahiers-nantais.fr/index.php?id=1235 [Accessed 22 November 2021].

Rolland, J. (2015). Le débat "Nantes, la Loire et nous". Nos engagements pour demain. Report, Nantes Métropole, Nantes.

Salles, D. (2006). *Les défis de l'environnement : démocratie et efficacité*. Syllepse, Paris.

de Toledo, C. (2021). *Le fleuve qui voulait écrire. Les auditions du parlement de Loire*. Manuella, Les liens qui libèrent, Paris.

PART 4

Anticipating the Future of Estuarine Cities

8

Past and Future Socio-Ecological Pathways of the Seine Estuary

An estuary is the place where a river meets the sea, resulting in a number of biophysical processes that characterize estuarine ecological functioning. However, the river drains a watershed occupied by a human society, for which the estuary also constitutes an obligatory passage when it exchanges with the ultra-marine periphery.

The example of the Seine estuary, and the major stages of its progressive artificialization since the middle of the 19th century, is particularly illustrative of this duality, typical of socio-ecosystems. We analyze it here according to the territorial ecology approach (Buclet 2015; Barles 2017) by examining the material flows associated with commercial traffic in the ports of Rouen and Le Havre. We will show how the history of the development of the estuary is conditioned by the development of a material demand generated by the development of the watershed, and more particularly of Paris.

The benefit of a long-term approach lies in a better capacity to predict the future. Therefore, two contrasting scenarios of the future of the socio-ecological system of the Seine will be sketched in the extension of our retrospective analysis in order to show the possible alternatives for the year 2050. It will be seen that the future of the estuary is closely linked to the future of the watershed, while being marked by the legacies of its past functioning, which afford it considerable inertia.

Chapter written by Gilles BILLEN, Julia LE NOË, Camille NOÛS and Josette GARNIER.

8.1. The Seine estuary as a socio-ecological system

A socio-ecological system is a hybrid object at the interface between nature and culture, constructed by the voluntary action of a society on the natural environments on which it depends. Estuaries, and the Seine in particular, are socio-ecosystems (Ostrom 2007, 2009) that are entirely emblematic of this duality. As a natural environment, the estuary constitutes an interface zone between the river and the sea, in which a number of biophysical processes that characterize estuarine ecological functioning take place: it is a zone of trapping, concentration and accumulation of numerous biogenic or mineral elements from the sea and the watershed. As a result, estuaries are hot spots of biodiversity and biological productivity: these environments are important for the nutrition and reproduction of many species of fish, birds and mammals, as is generally the case with *front* zones, at the meeting of different water masses or at the edge of contrasting aquatic or land environments. Up until the beginning of the 19th century, fishing was a major activity in the Seine estuary. Poles and stakes were permanently installed to fix wide-open nets in the direction of the current. Flatfish (sole, flounder), migratory fish (shad, lamprey, smelt, salmon, eel) and even cetaceans (blubber fish) were caught in abundance (Darsel 1972; Deschamps 2016).

Estuaries are also, by their nature of interface between land and sea, environments of very unstable morphology, subject to repeated and frequent submergence, as shown on the map of the course of the Seine from Le Havre to Pont-de-l'Arche (*BHF, fonds BB d'Anville*) drawn by the Magin brothers around 1740 (Roupsar 2008) (Figure 8.1).

This structural morphological instability of the estuaries is a serious obstacle to navigation. In the first half of the 19th century, it took eight days for ships to travel up the Seine to Rouen, a historic port city at the bottom of the estuary, and their draft could not exceed 3 m (Le Sueur 1989). From the middle of the 19th century, considerable work was undertaken to dam and stabilize the estuary channel, in parallel with the channeling of the Seine river axis to Paris. Created in 1530, the port of Le Havre, with direct access to the open sea, developed as a military stronghold and, at the same time, especially with the colonies, for the transport of passengers and as a commercial port for light materials. The digging of a canal linking it to Tancarville made it possible to avoid the difficult passage of the lower estuary. After World War I, the vast landing zones, constituted in the estuary on both sides of the stabilized and dredged navigation channel, allowed for the establishment of heavy industries, mainly petrochemicals. The port of Le Havre specialized in supplying these. Its complete destruction during World War II only temporarily interrupted this industrialization process, which culminated in the mid-1970s with the creation of the Antifer oil terminal. With Port 2000, a new set of facilities designed to accommodate the largest container ships was installed south of Le Havre, in the northern trench at the entrance to the estuary. The development of the outer harbor led to competition with the port of Rouen, whose accessibility must be constantly improved in order to ensure access for larger vessels. The opening of a new channel in 1961 ensured a minimum draught of 8 m, which was increased to 10 m in 1979 by the extension of the northern dike. Subsequent works guarantee a draft of 11.5 m today (Foussard et al. 2010). These stages in the development of the estuarine space, between the historic port at the bottom of the estuary and its outer harbor, are very similar to those followed by all the major ports on the Atlantic coast and the North Sea, such as Nantes–Saint-Nazaire, Bordeaux-Verdon, Antwerp-Zeebrugge, Bremen-Bremerhaven (Brocard et al. 1995; Lecoquierre 1999).

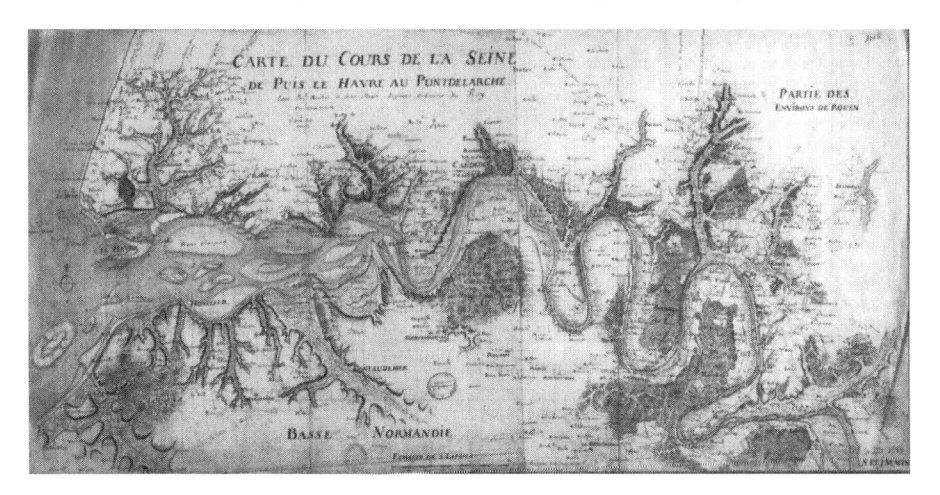

Figure 8.1. *Map showing the course of the Seine from Le Havre to Pont-de-l'Arche, drawn by the Magin brothers around 1740 (Roupsar 2008)*[1]. *For a color version of this figure, see www.iste.co.uk/salles/estuarinecities.zip*

---

1 https://www.persee.fr/doc/annor_0003-4134_2008_num_58_3_6207.

From the middle of the 19th to the end of the 20th century, navigation thus took precedence over all other functions of the estuary; it dictated its development in a heavy and irreversible manner (Figure 8.2).

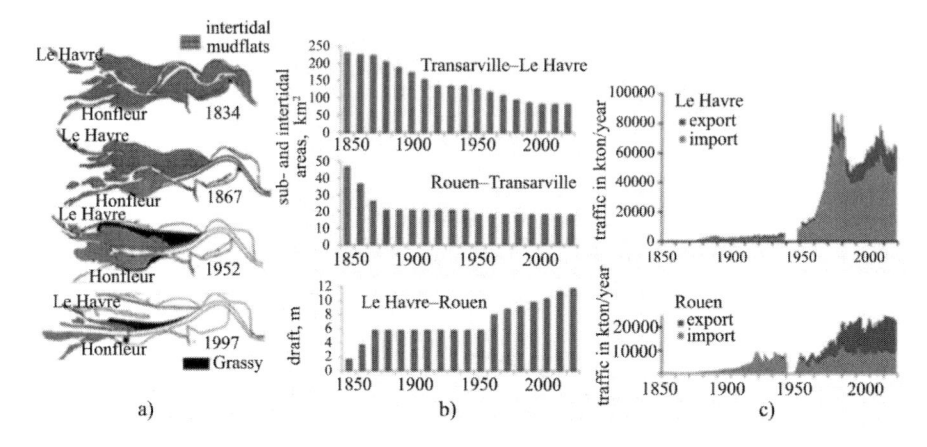

Figure 8.2. *Morphological transformation of the Seine estuary and evolution of maritime traffic in the ports of Le Havre and Rouen. For a color version of this figure, see www.iste.co.uk/salles/estuarinecities.zip*

COMMENTARY ON FIGURE 8.2.– *a) Morphological evolution of the Seine estuary since the mid-19th century (Avoine 1981; Foussard et al. 2010). b) Reduction of sub- and inter-tidal mudflats in the Tancarville-Le Havre and Rouen-Tancarville sectors, following the development of the navigation channel and the minimum draught of the navigation channel to Rouen (Foussard et al. 2010). c) Inbound and outbound traffic of the port of Le Havre and Rouen in thousands of tons per year (compilation of numerous sources, including Eurostat; the Ministry of Ecology; the Ministry of Transport; Morisot (1951); Grellet (1953); Malon (2006); Rai-Punsola (2010)).*

There is a close parallel between the major morphological changes imposed on the estuary and the development of traffic in both ports (Figure 8.2c). Although ecological compensation measures have been taken in the case of the most recent Port 2000 developments (construction of a dune resting place and an islet for seabirds off Honfleur, opening of a breach in the northern dike, digging of a meander in the large intertidal mudflats to slow down its deterioration, etc.), one can nevertheless measure the derisory character of these actions in the face of the profound mutation that has affected the estuary system over the last century and a half. The last professional fisherman working in the upper Seine estuary ceased their activities in the early 1980s, and the professional fleet working in the lower estuary was limited in 2010 to seven boats dedicated to eel and shrimp fishing (Morin 2010). Recreational fishing for fish and shellfish has been banned everywhere since 2008 due to environmental contamination (Fisson 2004).

8.2. The successive phases of port traffic

Since the transport function is at the root of the changes that the estuary system has undergone since the middle of the 19th century, it is essential, in order to understand the determinants, to analyze the content of port traffic and its evolution during this period (Figure 8.3).

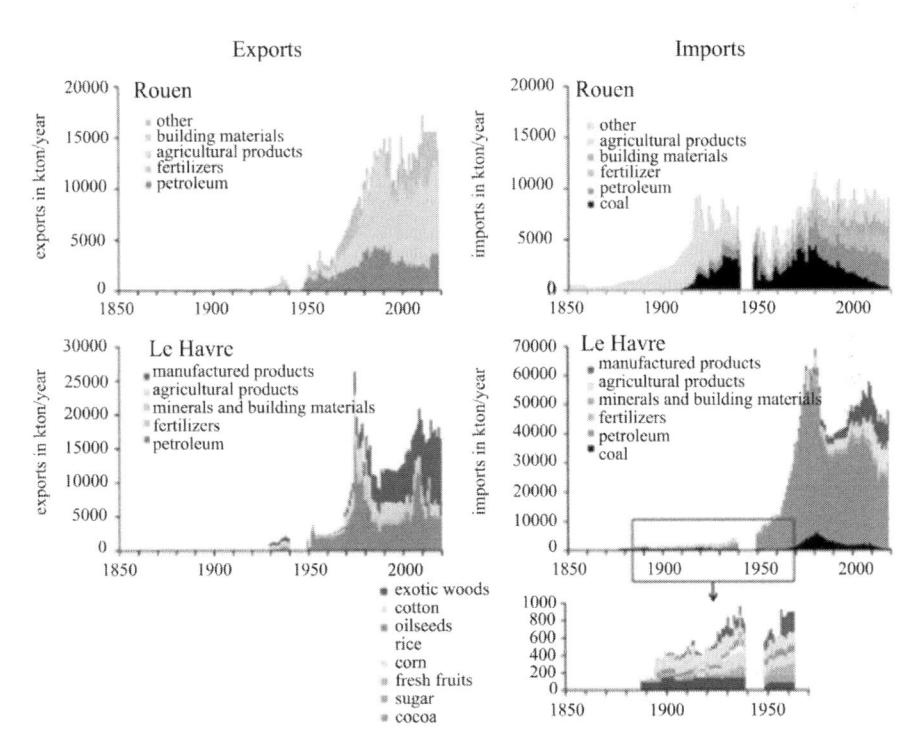

Figure 8.3. *Main goods making up the import and export traffic of the ports of Rouen and Le Havre since 1850 (compilation of numerous sources, including: Eurostat; the Ministry of Ecology; the Ministry of Transport; Morisot (1951); Grellet (1953); Malon (2006); Rai-Punsola (2010)). For a color version of this figure, see www.iste.co.uk/salles/estuarinecities.zip*

From 1900 to 1950, the ports of Le Havre and Rouen were mainly importers. Coal occupied a major place in the upstream traffic of Rouen, colonial products (coffee, cocoa, sugar, fruits, cotton, wood, etc.) fed the traffic of Le Havre. After World War II, petroleum products took over from coal in the import of fossil fuels, and the development of the port of Le Havre was largely devoted to handling petroleum. At the same time, Rouen became the primary port for the export of cereals, and fertilizers, mainly phosphates, took an increasingly prominent place in its imports. After the 1980s–1990s, petroleum imports were back, and the growth of the port of Le Havre was mainly linked to the growth of the containerized traffic of manufactured goods, both for import and export.

This review highlights three distinct phases in the development of the ports and the estuary: during the late 19th century and the first half of the 20th century, the great acceleration in the 30 years following World War II, and the post-1980s period, which is marked more by a stagnation in trade flows. The evolution of the composition of traffic that characterizes these three phases reflects three major sectors of economic activity in the territories served by the port: energy supply, the agri-food system and trade in manufactured goods.

8.3. The energy supply of the Seine basin

A detailed analysis of the evolution of energy consumption and supply in the Paris metropolitan area since the 18th century shows the gradual replacement of wood by coal as the main source of primary energy[2] during the 19th century; then the hegemony of oil in the first half of the 20th century, and finally – after the oil crisis of the 1970s – the rise of natural gas and nuclear electricity (Kim and Barles 2012). Total primary energy consumption per capita in the Paris metropolitan area appears to be slightly lower than the national average, but its evolution is quite similar, with a slow increase from the mid-19th century to the mid-20th century, followed by a tripling in price between 1950 and 1980, and a stabilization thereafter (Figure 8.4).

---

[2] That is, before the eventual conversion into energy carriers, such as fuels, electricity and gas distribution.

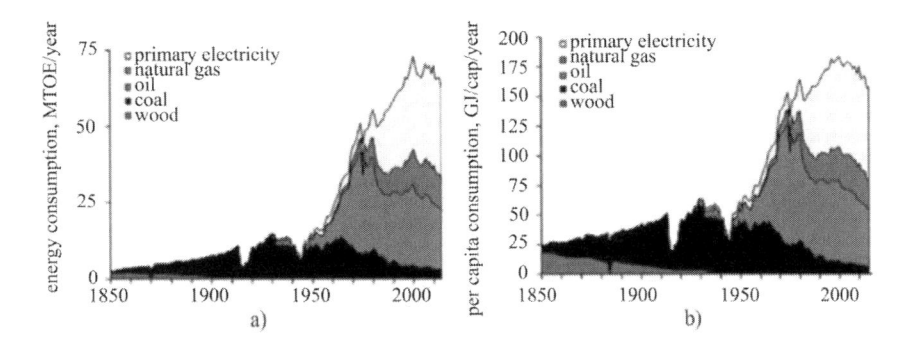

Figure 8.4. *Primary energy consumption in the Seine basin since the mid-19th century. a) In tons of oil equivalent (toe). b) In gigajoules (GJ) per capita per year (sources: Annuaires statistiques de la France, 1895–..., SOeS[3], Perron (1996); Kim and Barles (2012)). For a color version of this figure, see www.iste.co.uk/salles/ estuarinecities.zip*

The origin and modes of transport of the Ile-de-France energy supply have also been analyzed (Kim and Barles 2012). While wood was mainly transported by floating from the forested regions upstream of the basin, at the end of the 19th century coal was mainly transported by water (rivers and canals) from the collieries in the North, Belgium and then Lorraine, and to a very small extent by maritime import from England. The quantities of coal unloaded at the port of Rouen, which peaked at around 3,000 kilotons/year in the 1930s, still remained below 20–30% of the total consumption of the Seine basin.

From the 1920s onwards, maritime traffic became predominant for the supply of oil. At first, it was limited to the import of refined products, then it developed with the import of crude oil from 1929, and the commissioning of the Petit-Couronne refinery and, in 1933, with the launch of Notre-Dame-de-Gravenchon and Gonfreville-l'Orcher. At the same time as the port of Le Havre was being rebuilt after World War II, a pipeline system was set up linking the three refineries directly to a new oil terminal at Antifer. During the "Trente Glorieuses" (the "thirty glorious" years following World War II), oil traffic was the main driver of port development. However, since the 1980s, oil infrastructures have been operating at overcapacity. The Petit-Couronne refinery was permanently closed in 2012 (Tourret 2016).

---

3 http://developpement-durable.bsocom.fr/statistiques/.

The period following the two oil crises of the 1970s was marked by a diversification of energy supply, with the arrival of natural gas and nuclear power. Three nuclear power plants were commissioned on the Normandy coast between 1984 and 1990, with a total capacity of 10.4 gigawatts (GW), to which will be added the Flamanville EPR (1.7 GW), scheduled, after numerous delays, for 2023. Onshore wind farms in Normandy and Hauts-de-France have been developing 5 GW of power since 2020. Three large offshore wind farms are under construction and will add 1.7 GW of power by 2023. As we can see, the era of all-oil is over, and its replacement by both renewable and nuclear energy is well underway. The Normandy region remains well positioned as the energy supplier for the Seine basin (for which it already provides nearly a quarter of the current electricity consumption); nevertheless, the port infrastructures are becoming less crucial in this area.

8.4. The contribution of ports to the agri-food system of the Seine basin

Fertilizers are now the third largest item of import traffic in the port of Rouen (Figure 8.3). As early as 1907, large industrial capacities for the production of phosphate fertilizers (superphosphates) were developed in the Seine valley. Taking advantage of sulfuric acid production capacities, in existence since the end of the 18th century in the Rouen region (Fisson 2014), these industrial activities were fueled by phosphate ores imported for the most part from the French colonies or protectorates in North Africa (Figure 8.5) (Le Noë et al. 2020). Similarly, nitrogen fertilizer factories, using the Haber-Bosch process from atmospheric nitrogen, with the help of fossil fuels (coal, then natural gas; 1 ton of oil equivalent per ton of fixed nitrogen), were established in the Rouen region as early as 1930. On the one hand, the highly polluting nature of the phosphate industry, which produces 5 tons of phosphogypsum per ton of superphosphate contaminated with cadmium and uranium, and on the other hand, the accession to independence of the former colonies and protectorates led to the complete cessation of this activity in 1992. Today, already processed phosphate fertilizers are still imported, mainly from North Africa.

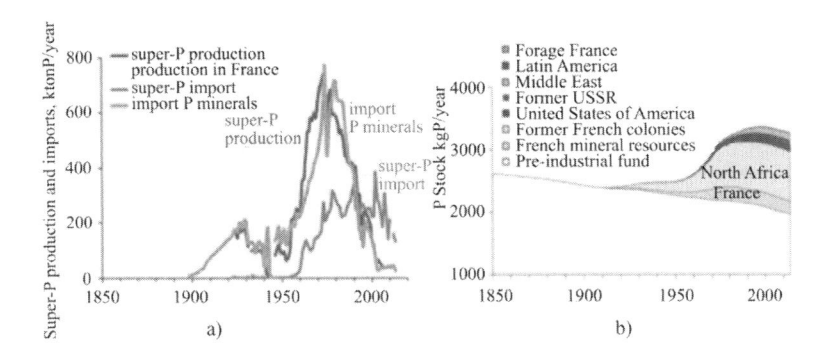

Figure 8.5. a) Production and importation of P ores and phosphate fertilizers in France. b) Increase in P content of arable soils in the Seine basin as a result of phosphorus fertilizer use during the 20th century (Le Noë et al. 2020). For a color version of this figure, see www.iste.co.uk/salles/estuarinecities.zip

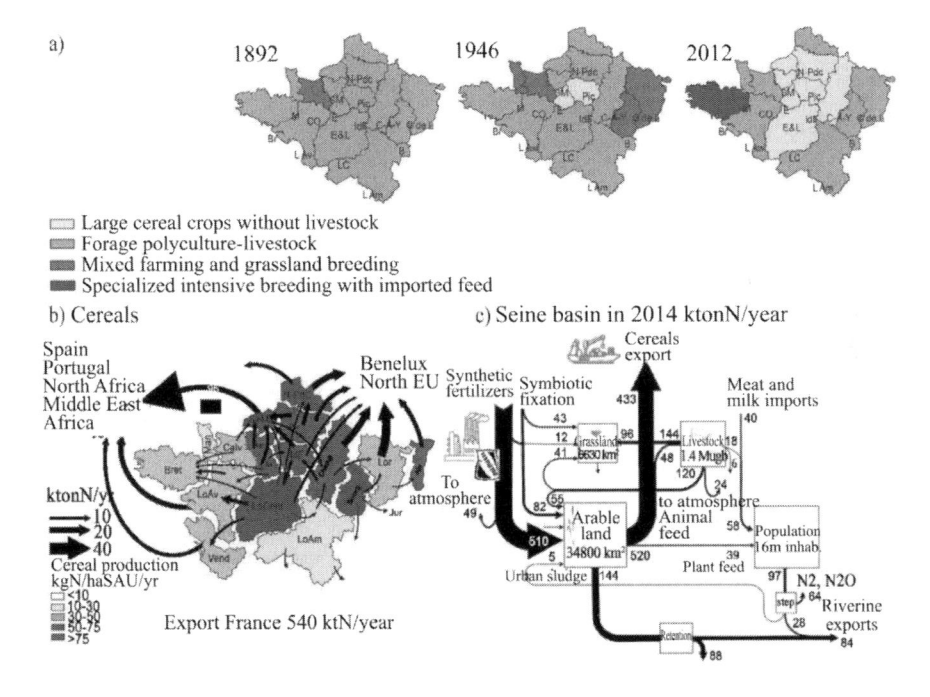

Figure 8.6. a) Territorial specialization of agricultural regions in northwestern France since the end of the 19th century. b) Main grain exchanges between the agricultural regions of the Seine basin and the Great West. c) Grain exports and imports in northern France (source: Le Noë et al. (2016, 2018)). For a color version of this figure, see www.iste.co.uk/salles/estuarinecities.zip

The massive use of industrial fertilizers, from the beginning of the 20th century but especially after World War II, made possible not only the intensification of agricultural production, but also a process of extreme territorial specialization by which the mixed farming-livestock system, dominant throughout France until the middle of the 20th century, gave way to large-scale cereal farming in the center of the Paris Basin. Livestock farming has been pushed back to the east, in regions that remain focused on polyculture-livestock farming, and to the west, where a specialized livestock farming system has been established, characterized by a sharp increase in livestock density and increased dependence on imports for livestock feed (Figure 8.6) (Le Noë et al. 2018).

Cereal production in the Paris Basin then became very much in surplus in relation to human and animal consumption in the same area (Figure 8.7a). From 1980 onwards, these surpluses constituted the main export traffic item for the port of Rouen (Figure 8.3): half of French wheat and barley exports and 20% of European cereal exports passed through the port of Rouen[4]. Eight silos were built around Rouen between 1960 and 1990, and a ninth was constructed in 2015, bringing the grain storage capacity to 1.3 million tons. These huge silos also profoundly marked the landscape of the Seine Valley and the petrochemical industry. Most grain exports are currently destined for North Africa and the Middle East (Figure 8.7b). This shift from a largely self-consuming and nourishing agriculture to one dominated by exports completes the integration of agriculture into the global market economy in which land and its products have become commodities, just like any other (Polanyi 1944). Until then, a series of protective measures – initiated by Jules Méline in the 1890s, completed with the establishment of the *Office national interprofessionnel du blé* (National Interprofessional Wheat Office) by the *Front Populaire*, and extended by the Monnet Plan in 1945, followed by the creation of the common agricultural policy (CAP) in 1962 (Duby and Wallon 1978, 1993) – had effectively protected French, and then European agriculture from integration into international markets. Inversely, the reforms of the CAP initiated in the early 1990s, marked a liberal turn towards an ever-increasing integration of French agriculture into global markets (Kroll and Pouch 2012; Plihon 2012; Bureau and Thoyer 2014). The transformations observed in the traffic of the port of Rouen, affecting the morphology of the estuary, are clearly the result of this set of supranational political and economic mutations.

---

4 www.haropaports.com/fr/rouen/produits-agroalimentaires.

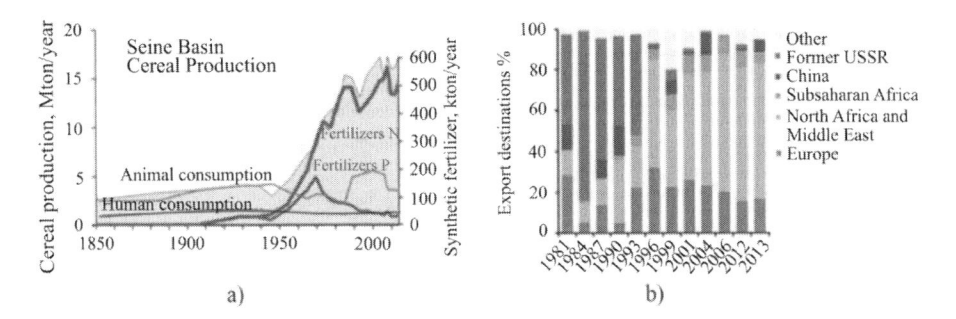

Figure 8.7. a) Cereal production, human and animal consumption of cereals, and use of industrial fertilizers (N and P) in the Seine basin since 1850 (source: Le Noë et al. (2018)). b) Destination of cereal exports from Rouen (source: SitraM database on commodity flows). For a color version of this figure, see www.iste.co.uk/salles/estuarinecities.zip

From the 1990s onwards, European public policies, which had become less interventionist in terms of encouraging an increase in agricultural productivity since the end of the 1980s, moved towards more environmental regulation. This has resulted in a slowing down of intensification (reduction of phosphorus iron fertilization, capping of nitrogen fertilization), without hindering the increase in the use of pesticides and the accentuation of the opening and territorial specialization of agriculture. Coastal contamination by eutrophication, through the proliferation of unwanted green algae, is a persistent environmental problem (Garnier et al. 2019a).

8.5. The era of globalized trade in manufactured goods

The invention and standardization in the early 1960s of the "container", a metal box that can be easily loaded and stowed in the hold or on the deck of a ship, on a barge, on a truck or on a train, revolutionized the maritime and inland transportation of manufactured goods. The cost of long-distance transport has been reduced to an unprecedented degree, becoming a negligible part of the value of the goods transported. The logistics of "containerization" has thus made possible the growth of trade in manufactured goods around the world (Aubin 2000; Frémont 2005). To accompany this revolution, the port of Le Havre built its first two container terminals in the inner basin of the port in 1968 and 1972. Smaller container unloading sites were also developed by the port of Rouen in 1977 and 1978 at Grand-Couronne and Radicatel. In order to keep up with the evolution of the size of cargoes, the Le Havre terminals were moved to the tidal basin in 1992–1995, before a new deep-water extension of the port in the mouth of the Seine (Port 2000) was built in 2000–2010.

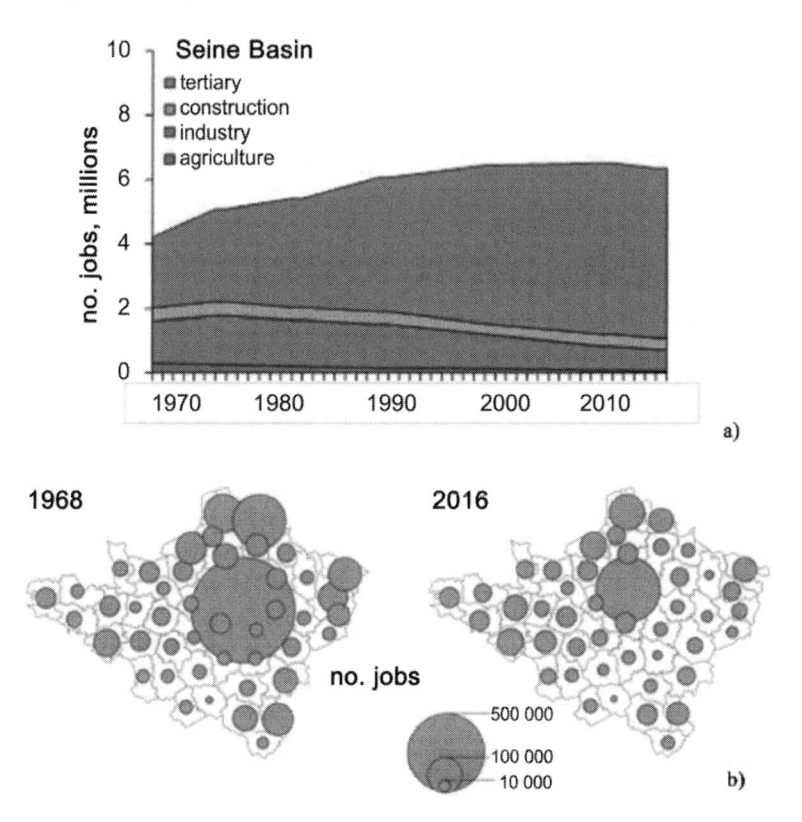

Figure 8.8. *a) Changes in employment in the Seine basin from 1968 to 2016. b) Geographical distribution of industrial employment in 1968 and 2016 (source: Insee, population census: active population aged 25–54 years at the place of work). For a color version of this figure, see www.iste.co.uk/salles/estuarinecities.zip*

Despite these massive investments in very heavy infrastructures, Le Havre only ranks fifth for container traffic of the ports of the "North European row", with 23 Mton in 2012, behind Rotterdam (125 Mton), Antwerp (104 Mton), Hamburg (89 Mton) and Bremerhaven (65 Mton). For the 10 major companies that handle 70% of the world's container traffic in 2018 (UNCTAD 2018), the North European row, which allows for the service of the entire northwestern part of Europe, is considered a single entity for exchange with the ports of the eastern seaboard of the United

States and Asia. And the logic of the choice of port of call within this large continental grouping pays little attention to the particularities of the countryside of each port, or to the precise final destination of the goods once unloaded. The extremely rapid expansion of the globalization of production and port trade in manufactured goods is thus going hand in hand with the deindustrialization and loss of industrial jobs in France in general, and in the Seine basin in particular (Figure 8.8).

8.6. What is the future of the Seine estuary?

The future of the Seine estuary is totally linked to that of the two major ports of Le Havre and Rouen, whose development closely follows the evolution of the metabolism of the upstream territories and the logic of international flows. In view of the retrospective analyses, we can imagine two contrasting trajectories for the future of the basin and the estuarine territory.

The first possible trajectory is the continuation of the major trends that have been described. In his report *Paris et la mer: la Seine est Capitale* (Paris and the Sea: The Seine is Capital), Jacques Attali (2010) lists 50 proposals to "make Greater Paris the ocean gateway and natural capital of Western Europe". For Attali, "the development of the Ile-de-France region will have to be driven by maritime trade from tomorrow onwards". Because, he added, "the densification of world trade, global demographics, and the massification of flows will leave people with a choice between museum-like coastal cities or ports that proudly stand out at sea with their industry" (*author's translation*). Nicolas Sarkozy's speech in Le Havre on the 16th of July 2009 was along these lines: "We must rebuild a maritime policy and ambition for France, based on the new challenges […] of a globalized planet that breathes through international trade" (*author's translation*).

These injunctions draw for the Seine basin a future of urban concentration on the Paris–Le Havre axis, of ever greater specialization of agricultural and industrial activities, of opening up to international markets. For the estuary, this development logic would imply the continuation of works intended to allow access to bulk carriers of more than 60,000 tons, and the realization of increasingly dense road, rail and waterway connections. The Seine estuary would remain condemned to being what it has been for more than half a century: a transport infrastructure, leaving little room for the rest of animal and landscape biodiversity.

In agriculture, a scenario of continued opening and increased specialization of the basin's agri-food system – built according to the logic of conventional agriculture, even though it respects current environmental rules – would allow for a doubling of French cereal exports, from 28 Mton/year today to 57 Mton in 2050 (Billen et al. 2018). This would, however, lead to a considerable increase (nearly a factor of 4 on a national scale) in feed imports in regions specialized in intensive livestock production. In addition, this scenario would lead to a significant worsening of nitrate and pesticide contamination of water and a 45% increase in greenhouse gas emissions from the agricultural sector (Garnier et al. 2019b).

An alternative scenario, more in line with the commitments made by France to achieve carbon neutrality by the middle of the century (*Ministère de la Transition écologique et solidaire 2020*: Ministry for ecology and solidarity transition 2020), can be imagined for the Seine estuary. Such an objective presupposes the systematic substitution of the uptake of fossil and fissile fuels by renewable, solar, wind and biomass energies. In this energy scenario, described in detail by the negaWatt[5] association, petroleum-based infrastructures become obsolete. As far as agriculture is concerned, the Afterres2050 scenario (Solagro 2016), and also the scenario of autonomy, reconnection and decrease (ARD in French) for the proportion of animal proteins in the diet (ARD, Billen et al. 2018), show that it is perfectly possible to feed France, in a healthier way than today, by reducing industrial fertilizers by 60% (Afterres 2050) or by doing without them completely (ARD). These scenarios predict the maintenance of cereal exports at around 40% of the current level to regions of the world that still have a deficit. This scenario of energy and agricultural evolution shows that water quality would be greatly improved (Garnier et al. 2019a), and greenhouse gas emissions reduced by a factor of two (Garnier et al. 2019b). Finally, in line with these alternative scenarios, greater sobriety in the consumption of manufactured objects, an increase in their lifespan and their recycling capacity, combined with the relocation of their production, would considerably reduce the need for long-distance transport of consumer goods and our dependence on the importation of these objects.

---

5 https://negawatt.org/Scenario-negaWatt-2017-2050.

The reduction of pressures on the estuary could thus make it possible to consider the recovery of certain ecological functions (water quality, biodiversity and biological production), even though the legacy of the past trajectory of the estuary, deeply rooted in its own morphology, makes it unlikely to be able to recover all of the lost functions, such as those of the wetlands and the large intertidal mudflats.

8.7. Conclusion

The approach adopted is based on the historical analysis of the material flows that characterize the metabolism of the territories. It has shown how closely the evolution of the Seine estuary has been linked, for more than a century and a half, to that of the socio-ecosystem of its upstream basin.

A flow analysis reveals an undeniable material reality, and thus renews the classical economic approach that quantifies exchanges in terms of monetary value, without always questioning what this value represents (a scarcity, a quantity of mobilized labor, a balance of power), or especially the purpose of these exchanges, whose increase is often considered as an end in and of itself.

It is therefore clear that the development of the Seine estuary responds to the growing centralization and urban polarization of Paris: to maintain its position as a world city in an increasingly globalized economy, Paris must be linked to the rest of the world.

The continuation of the major historical trends in the evolution of the Seine basin (opening up of material cycles, agricultural specialization, desertification of the rural regions of the eastern part of the basin, hyper-development of the Paris–Le Havre axis) would inevitably lead to the further artificialization of the estuary. The progressive loss of its remaining ecological functions would be inescapable, despite the compensatory measures that could be taken (de Billy et al. 2015; Weissgerber et al. 2019).

However, it is possible to imagine a breakthrough scenario, where certain trends or new aspirations that are emerging through many social movements would be pushed to the extreme: re-localization of supplies, decentralization and revitalization of peripheral regions, development of agroecological practices, a reduction in the share of animal proteins in the diet. Such a scenario would make it possible to considerably reduce the pressures on the estuary, and to imagine a polyfunctional future for it.

The territorial metabolism approach used here highlights the field of biophysical possibilities. It opens the way to more creative forward thinking, which could collectively involve all stakeholders, from managers to communities, from researchers to associations. The options for the future would no longer be constrained by a pre-existing economic framework; the economy would regain its rightful place at the service of a true social project, rather than as a single horizon.

8.8. References

Attali, J. (2010). *Paris et la mer : la Seine est Capitale*. Fayard, Paris.

Aubin, C. (2000). Stratégie des firmes et échanges internationaux. *Le commerce mondial. Cahiers français*, 299, 26–33.

Barles, S. (2017). Écologie territoriale et métabolisme urbain : quelques enjeux de la transition socioécologique. *Revue d'économie régionale & urbaine*, 5, 819–836. doi: doi.org/10.3917/reru.175.0819.

Billen, G., Le Noë, J., Garnier, J. (2018). Two contrasted future scenarios for the French agro-food system. *Science of the Total Environment*, 637–638, 695–705. doi: 10.1016/j.scitotenv.2018.05.043.

de Billy, V., Tournebize, J., Barnaud, G., Benoît, M., Birgand, F., Garnier, J., Lesaffre, B., Lévêque, C., de Marsily, G., Muller, S. et al. (2015). Compenser la destruction de zones humides. Retours d'expérience sur les méthodes et réflexions inspirées par le projet d'aéroport de Notre-Dame-des-Landes (France). *Natures Sciences Sociétés*, 23, 27–41. doi: 10.1051/nss/2015008.

Brocard, M., Lecoquierre, B., Mallet, P. (1995). Le chorotype de l'estuaire européen. *Mappemonde*, 3, 6–7.

Buclet, N. (2015). *Essai d'écologie territoriale : l'exemple d'Aussois en Savoie*. CNRS, Paris.

Bureau, J.-C. and Thoyer, S. (2014). *La politique agricole commune*. La Découverte, Paris.

Charlier, J. (1990). L'arrière-pays national du port du Havre. *Espace géographique*, 19/20, 325–334. doi: doi.org/10.3406/spgeo.1990.3014.

CNUCED (2018). Étude sur les transports maritimes. *Organisation des Nations Unies* [Online]. Available at: https://unctad.org/fr/PublicationsLibrary/rmt2018_fr.pdf.

Darsel, J. (1972). L'Amirauté en Normandie : amirauté de Caudebec-Quillebeuf, 2e partie. *Annales de Normandie*, 22, 105–131 [Online]. Available at: https://www.persee.fr/collection/annor.

Deschamps, G. (2016). *La pêche à pied : histoire et techniques*. Quae, Paris.

Duby, G. and Wallon, A. (1978). *Histoire de la France rurale, tome IV : Depuis 1914*. Le Seuil, Paris.

Duby, G. and Wallon, A. (1993). *Histoire de la France rurale, tome III : Apogée et crise de la civilisation paysanne (de 1789 à 1914)*. Le Seuil, Paris.

Fisson, C. (2014). L'estuaire de la Seine : état de santé et évolution. Report, Fascicule Seine-Aval 3.1, GIP Seine-aval, Rouen.

Foussard, V., Cuvilliez, A., Fajon, P., Fisson, C., Lesueur, P., Macur, O. (2010). Évolution morphologique d'un estuaire anthropisé de 1800 à nos jours. Report, Fascicule Seine-Aval 2.3, GIP Seine-Aval, Rouen.

Frémont, A. (2005). Conteneurisation et mondialisation. Les logiques des armements de lignes régulières. PhD Thesis, Université Panthéon-Sorbonne – Paris I, Paris.

Garnier, J., Billen, G., Legendre, R., Riou, P., Cugier, P., Schapira, M., Théry, S., Thieu, V., Menesguen, A. (2019a). Managing the agri-food system of watersheds to combat coastal Eutrophication: A land-to-sea modelling approach to the French coastal English channel. *Geosciences*, 9, 441. doi:10.3390/geosciences9100441.

Garnier, J., Le Noë, J., Marescaux, A., Sanz-Cobena, A., Lassaletta, L., Silvestre, M., Thieu, V., Billen, G. (2019b). Long-term changes in greenhouse gas emissions from French agriculture and livestock (1852–2014): From traditional agriculture to conventional intensive systems. *Science of the Total Environment*, 660, 1486–1501. doi.org/10.1016/j.scitotenv.2019.01.048.

Grellet, H. (1953). Rouen Grand port moderne V – Activité du port, analyse des principaux trafics. *Études normandes*, 7, 389–428. doi.org/10.3406/etnor.1953.3068.

Kim, E. and Barles, S. (2012). The energy consumption of Paris and its supply areas from the 18th century to the present. *Regional Environmental Change*, 12, 295–310. doi:10.1007/s10113-011-0275-0.

Kroll, J.-C. and Pouch, T. (2012). Régulation versus dérégulation des marches agricoles : la construction sociale d'un clivage économique. *L'Homme & la société*, 183–184, 181–206. doi: 10.3917/lhs.183.0181.

Le Noë, J., Billen, G., Lassaletta, L., Silvestre, M., Garnier, J. (2016). La place du transport de denrées agricoles dans le cycle biogéochimique de l'azote en France : un aspect de la spécialisation des territoires. *Cahiers Agricultures*, 25, 15004. doi: 10.1051/cagri/2016002.

Le Noë, J., Billen, G., Esculier, F., Garnier, J. (2018). Long-term socioecological trajectories of agro-food systems revealed by N and P flows in French regions from 1852 to 2014. *Agriculture, Ecosystems & Environment*, 265, 132–143. https://doi.org/10.1016/j.jenvman.2017.09.039.

Le Noë, J., Roux, N., Billen, G., Gingrich, S., Erb, K.-H., Krausmann, F., Thieu, V., Silvestre, M., Garnier, J. (2020). The phosphorus legacy offers opportunities for agro-ecological transition (France 1850–2075). *Environmental Research Letters*, 15, 064022. doi: doi.org/10.1088/1748-9326/ab82cc.

Le Sueur, B. (1989). La navigation en Basse-Seine au début du XIXe siècle. *Cahier du musée de la batellerie*, 25.

Lecoquierre, B. (1999). Un modèle diachronique des ports de la Basse-Seine dans l'espace européen. In *Les Ports normands : un modèle ?* Wauters, E (ed.). Presses universitaires de Rouen et du Havre, Mont-Saint-Aignan.

Malon, C. (2006). *Le Havre colonial de 1880 à 1960*. Presses universitaires de Rouen et du Havre/Presses Universitaires de Caen, Mont-Saint-Aignan/Caen.

Ministère de l'Écologie (2010). Bilan annuel des ports maritimes et voies navigables [Online]. Available at: www.developpement-durable.gouv.fr.

Ministère de l'Écologie (2020). Stratégie française pour l'énergie et le Climat. Programmation pluriannuelle de l'énergie 2019–2023, 2024–2028 [Online]. Available at: https://www.ecologique-solidaire.gouv.fr/sites/default/files/20200 422%20Programmation%20pluriannuelle%20de%20l%27e%CC%81nergie.pdf.

Ministère des Transports (1982). Annuaire statistique des transports [Online]. Available at: https://temis.documentation.developpement-durable.gouv.fr/pj/992/ 992_1981_1.pdf.

Morin, J. (2010). Poissons, habitats et ressources halieutiques : cas de l'estuaire de la Seine. Working document, Fascicule Seine-Aval, GIP-Seine-Aval, Rouen.

Morisot, J. (1951). Le complexe portuaire de la Seine Maritime. Rouen et Le Havre unis au service de l'Economie européenne. *Études normandes*, 1, 1–12. doi: doi.org/10.3406/etnor.1951.1274.

Ostrom, E. (2007). A diagnostic approach for going beyond panaceas. *Proceedings of the National Academy of Sciences of the United States of America*, 104(39), 15181–15187. doi: doi.org/10.1073/pnas.0702288104.

Ostrom, E. (2009). A general framework for analyzing sustainability of social-ecological systems. *Science*, 325, 419–422. doi: doi.org/10.1126/science. 1172133.

Perron, R. (1996). *Le marché du charbon, un enjeu entre l'Europe et les États-Unis de 1945 à 1958*. Éditions de la Sorbonne, Paris.

Plihon, D. (2012). *Le nouveau capitalisme*, 4th edition. La Découverte, Paris.

Polanyi, K. (1944). *La Grande Transformation. Aux Origines politiques et économiques de notre temps*. Gallimard, Paris.

Rai-Punsola, V. (2010). Les flux logistiques en Haute-Normandie. *Logistique Seine Normandie* [Online]. Available at: www.logistique-seine-normandie.com.

Reymondier, P. (1990). Les accès du port de Rouen (depuis l'origine jusqu'à l'époque actuelle). Synthèse pour le port de Rouen. Working document, GIP-Seine-Aval, Rouen.

Roupsar, M. (2008). Nicolas et Jean Magin, cartographes des côtes de la Manche au début du XVIIIe siècle. Essai d'inventaire de leur production [Online]. Available at: https://www.persee.fr/doc/annor_0003-4134_2008_num_58_3_6207.

Solagro (2016). Le scénario Afterres 2050 [Online]. Available at: https://afterres2050.solagro.org/wp-content/uploads/2015/11/Solagro_afterres2050-v2-web.pdf.

Tourret, P. (2016). Le Havre et Rouen : décryptage d'une dynamique portuaire. Working paper, ISEMAR synthesis note, 185.

Weissgerber, M., Roturier, S., Julliard, R., Guillet, F. (2019). Biodiversity offsetting: Certainty of the net loss but uncertainty of the net gain. *Biological Conservation*, 237, 200–208. doi.org/10.1016/j.biocon.2019.06.036.

9

Metropolitan Trajectories for Anticipatory Governance of Urban Biodiversity

The environmental policies of large cities, which are part of complex local geopolitical issues, are subject to modes of governance traditionally oriented towards attractiveness and urban and economic development. At the same time, the conservation of biodiversity has progressively become a public issue and now concerns all cities and their urbanization dynamics. The issue of biodiversity is in fact tending to be imposed, on the one hand, in urban governance by techniques and tools contributing to its legal inscription in operational framework documents (green and blue framework, urban planning documents, etc.), and, on the other hand, as a political subject integrated into other dimensions of this governance (quality of life, risks, mobility, etc.). The governance of biodiversity is therefore now at the heart of tensions and negotiations between different institutions (local authorities, government departments, real estate developments, business parks), and more specifically between departments dedicated to urban planning, land use, economic development and the environment. Given this observation, this chapter proposes to explore the way in which the issue of biodiversity is linked to urban policies, in order to think about the future of Bordeaux as an estuarine city.

While the search for a compromise between urban development and biodiversity conservation is hardly on the political agenda, the Nature

---

Chapter written by Charles DE GODOY LESKI and Yohan SAHRAOUI.

Department of Bordeaux Métropole launched a scientific and strategic prospective on this issue in 2017. Within the framework of the BiodiverCité research–action project, a community of stakeholders from various backgrounds was brought together around a common issue: urban biodiversity. BiodiverCité has been a unique experience of cooperation between researchers (sociology, geography, ecology, economics) and practitioners of the metropolitan area (experts, urban planners, territorial agents, associations, etc.), to work together to make visible and cognitively appropriate the socio-ecological interdependencies related to the preservation of urban biodiversity.

This multidisciplinary and multi-stakeholder cooperation resulted in a collaborative prospective, the final objective of which was to develop scenarios for metropolitan biodiversity by 2035. Five singular metropolitan trajectories were co-constructed in order to best anticipate a future that integrates the challenges of urban development and biodiversity preservation.

9.1. The challenges of an attractive city faced with ecological injunctions: contextual elements of emerging governance

As in all estuarine cities, the environmental issues of Bordeaux Métropole, such as the preservation of biodiversity, go beyond its institutional territory, that is its 28 communes, to extend to the estuarine functional territories (185 communes located in the Gironde and Charente-Maritime). These territories participate in the economic and social maintenance of the city (Ascher 1995) through productive sectors (wine, forestry, industry, etc.), as well as through recreational activities that are sometimes intertwined (fishing, hunting, etc.). In an ecological sense, they contribute to rendering a certain number of services to the city (flood expansion zones, cooling islands, limitation of erosion and the advance of sand dunes, etc.): the function is translated here by a service rendered by nature to urban dynamics. In terms of biodiversity, the ecological processes, linked in particular to the flow of materials and animals and plant species, therefore require us to think about this issue by again questioning asymmetrical interdependencies of centers and peripheries through this socio-ecological dimension. This overcoming of metropolitan boundaries and ordinary urban functioning, illuminated by the concept of the *metapole*

(Ascher 1995), involves a broader reflection on the ecosystems near and far from the metropolitan core.

Since the turn of the millennium, Bordeaux Métropole has displayed strong ambitions on several fronts: demographic, economic, scientific and mobility development. The political slogan of the 2010s, "Bordeaux as a millionaire city", initially aimed at 2030 (then 2050), and the metropolitan management of economic and tourist flows, have guided dense urbanization policies that question the robustness of administrative limits in contemporary public action. Due to its heritage and geographic location, the city benefits from a very successful image, reinforced by the classification of Bordeaux as a UNESCO World Heritage Site on June 28, 2007[1]. Within the framework of the OIN, Opération d'Intérêt National (national interest operation) Bordeaux Euratlantique, the arrival of the high-speed railway (HSR) linking Paris to Bordeaux as of July 2017, in just two hours, has structured strategic axes of national and European north–south mobility, and has stimulated the reclamation of industrial wastelands, particularly port wastelands, along the Garonne River[2]. The patrimonialization of the city of Bordeaux favors old and new buildings (Cité du Vin, the former Niel barracks, the Euratlantique business district, the Matmut-Atlantique stadium, the development of the Bassins à flot – Bacalan district) against attempts to direct the city's development through spaces of low economic value (marshes, wild banks), which are nevertheless necessary to maintain the urban and ecological functions of a "*water city*". Peri-urban and rural areas are faced with competition between the preservation of their forest biodiversity and the commodification of their land for the residential economy. The issue of biodiversity is thus balanced between the supply of and demand for

---

1 Bordeaux is the first urban assembly on such a vast and complex scale to be distinguished by the UNESCO World Heritage Commission since its creation in 1976. The perimeter inscribed on the World Heritage List includes almost all of Bordeaux within the boulevards, with the exception of the Belcier district located beyond the Saint-Jean train station.

2 Around the World Heritage perimeter, a heritage attention zone, or buffer zone, ensures the link between the city center and the bordering communes. This area includes the entire right bank of Bordeaux, that is, the La Bastide district and part of the communes of Floirac, Cenon and Lormont, which, over more than 500 hectares in area, offer a largely planted landscape. On the left bank, the buffer zone follows the route of the ring railroad. It includes elements of great interest, such as the Parc Bordelais or the Lescure district. On this side of the Garonne, the zone of heritage attention concerns, outside Bordeaux, the municipalities of Talence, Pessac, Mérignac, Bouscat and Bruges.

environmental goods and services, which constitute a quality living environment.

At the turn of the millennium, French cities began to integrate elements of "nature" into their territorial strategies and urban projects, notably under joint European or national regulatory injunctions: the DCE (2000 and 2011)[3], the Grenelle laws (2007 and 2010), the law for the reconquest of biodiversity and landscapes (2016), the GEMAPI reform of the MATPAM law (2014) and the NOTRe law (2015). The legislative injunctions are expressed locally in the codes of the environment, urban planning and construction and housing in particular. Progressively, the strategic appropriation of an environmental public action (Salles 2006; Lascoumes 2012) has been based on the landscape approach as a vector for the enhancement of agricultural and semi-natural spaces in service of metropolitan economic influence, in the case of the Bordeaux Métropole (Touchard 2019).

Between 2007 and 2011, the socialist Vincent Feltesse, president of the CUB, Communauté urbaine de Bordeaux, ex-Bordeaux Métropole (Bordeaux Urban Community, formerly Bordeaux Métropole) and Alain Juppé, mayor of Bordeaux (Les Républicains), launched the "Bordeaux Métropole 3.0" project, convening several stakeholders around the future of the city (municipalities, civil society, economic stakeholders and institutional partners) under the framework of a major participatory project. This was followed by the production of an important document for the community reference system: the programmatic document "*5 sens pour un Bordeaux métropolitain*" ("5 directions for a metropolitan Bordeaux", Bordeaux Métropole 2011). The Bordeaux Métropole urban project, through the major metropolitan projects, displayed the ambition to harmoniously articulate the demographic growth and urban densification through the implementation of OIN and OIM strategies, with the maintenance and enhancement of selected natural and semi-natural spaces. This is the first time in Bordeaux's local political history that ecological imperatives have been explicitly called upon in a large-scale urban project. The guiding principle of this urban project was to limit urban sprawl by encouraging the

---

3 The DCE, Directive-Cadre sur l'Eau (Water Framework Directive) requires monitoring of the ecological status of all the waters of the European Union. The normative pressure in the field of water, already present through numerous European directives on water quality and risks, is asserting itself as one of the strong constraints for urban development.

densification necessary to achieve the goal of a "millionaire city". The functions of nature in the urban configuration revealed the interdependencies between the social, economic and environmental dimensions. The principles of sustainable development have thus been applied in the "most diluted agglomeration in France" (A'urba 2021. *Construction et PLU 3.1. Bilans communaux et enseignements métropolitains*. Bordeaux Métropole urban planning agency, former director, *author's translation*), an argument that reunites urban planning with politics in an effort to strengthen metropolitan appeal.

In 2012, the political slogan "Bordeaux, a millionaire city", which was translated into urban project mode with the slogan "50,000 housing units around public transport", and renamed "50,000 housing units accessible through nature" in 2020, was supplemented by another slogan, "55,000 hectares of nature for the city", which drew on the "nature" frame of reference for the arguments of a political narrative that would help to balance the image of a booming urban area. Several observations can be made in reality and with a few years' hindsight: on the one hand, despite the political announcement of 55,000 ha for nature, the conceptual model for enhancing natural and semi-natural spaces is still largely invisible to citizens; on the other hand, the deployment of the "50,000 housing units around public transport" project is far from being achieved (Bordeaux and Métropole 2013a). This puts into perspective the image of a controlled expansion strategy and an anticipated contractualization with developers and lessors (Pinson et al. 2018). In this vast urban project, we find the classic tensions of the urban configuration (Lussault 2017) between the construction of heavy infrastructure (tramways, roads, networks) and its corollary in terms of the fragmentation of natural spaces.

In major urban projects, the preservation of biodiversity has most often taken place in the interstitial spaces left vacant by the planned urban operations. The launch in 2012 of an Atlas of metropolitan biodiversity for the general public (Bordeaux and Métropole 2016)[4] marks a desire on the part of the *direction de la Nature* (Nature Department) to acquire knowledge

---

4 This atlas is also available as a professional version, which compiles 29 technical notebooks listing seven types of ecological habitats (alluvial plain of the banks of the Garonne, limestone slopes of the right bank, Landes plateau, northern marshes, waterways, parks and green spaces, and buildings), with 551 species of fauna and 1,285 species of plants inhabiting these ecosystems.

in order to influence future negotiations with the *direction de l'Urbanisme* (Urban Planning Department). The strategy of carrying out an inventory of living things on a communal and then metropolitan scale was explicitly aimed at addressing elected officials as bearers of both communal and potentially metropolitan interests.

In the set of norms and practices of biodiversity governance, project management is confronted with increasing regulatory constraints. The ERC, "*Éviter, Réduire, Compenser*" (Avoid, Reduce, Compensate), doctrine is supposed to provide a more ecological legal framework for major development operations. The reduction of the anthropic impacts of an urban project is based on the logic of compensation in the wetlands of the fluvio-estuarine space. This dynamic of expanding metropolitan interests towards the estuary is itself a function of the capacity of the Bordeaux Métropole to concentrate human, technical and natural resources according to the development objectives formalized over the last 20 years: millionaire city, demographic and economic attractiveness and concentration of investments by the OIN and the OIM.

Bordeaux Métropole's ambition to reconcile territorial attractiveness, urban development and the preservation of biodiversity has been progressively translated into urban planning documents (PLUi, local inter-municipal urban plan, 2014) defining land use perspectives. The *direction de la Nature* (Nature Department) estimates that nearly 3,400 hectares of potential wetlands, 700 hectares of forests or 1,700 hectares of moorland and grasslands are not subject to any form of protection under PLUi. In this context, the BiodiverCité project, initiated by the *direction de la Nature* (Nature Department) of Bordeaux Métropole, is part of a strategy to preserve biodiversity in the face of urban development.

9.2. The cognitive stakes of a collaborative territorial prospective

The challenges of preserving biodiversity are characterized in particular by the complexity of the knowledge to be made coherent and then operationalized, both technically and politically, to adapt collectively to global changes. The BiodiverCité collaborative prospective approach has endeavored to make visible the effects of urbanization on the state of environments, by emphasizing the socio-political relationships that govern them (Leroy 2001; Jollivet and Pena-Vega 2002). We describe here the stage of co-construction of a common knowledge base among the stakeholders on the socio-ecological functioning of the Bordeaux Metropolitan area. The stakeholders involved in the participatory process were selected according to their membership in different structures acting in the fields of land use planning and nature conservation[5], and whose field of competence applies to the metropolitan territory, to its physical or administrative sub-entities or to a larger scale (department or region, for example).

The ARDI, *acteurs, ressources, dynamiques, interactions* (stakeholders, resources, dynamics, interactions) method (Étienne et al. 2011) was mobilized to co-construct, during a participatory workshop, a cognitive model of the urban socio-ecological system, understood as a system intertwining socio-political and ecological processes through, in particular, modes of governance by public action instruments (Schewenius et al. 2014).

In order to do this, participants were asked to list all the stakeholders identified in the system, the resources they use and their dynamics, as well as the interactions between the stakeholders and these resources. The synthesis diagram (Figure 9.1) was finalized by the researchers from the different contents (recorded discussions, note-taking and graphic diagrams) of the workshops, and then presented to the participants for correction and validation.

The territory's resources are made up of four categories:

– resources related to land use and land cover;

– primary resources, such as air, water, soil, groundwater, fauna and flora;

– public action instruments, such as zoning and urban planning and development documents (e.g. PLUi), biodiversity preservation or inventory (e.g. ZNIEFF), environmental and risk management (e.g. PPRI), ecological compensation systems and zones and financial instruments (e.g. taxes);

– intangible resources such as quality of life, territorial attractiveness, collective intelligence, education and training.

---

5 In total, 22 stakeholders were mobilized for this approach, from local government departments and mixed unions, State departments, naturalist and environmental associations, engineering firms, urban planning agencies, research laboratories, etc.

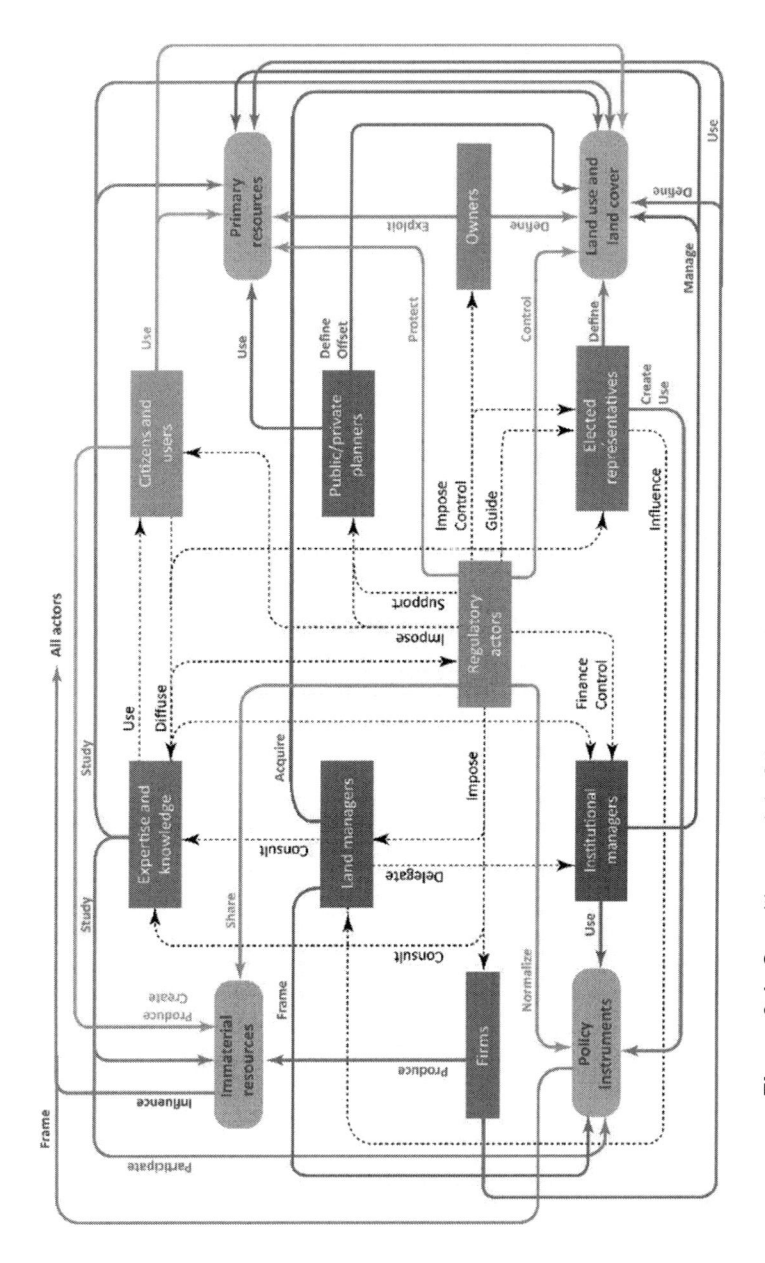

Figure 9.1. *Cognitive model of the metropolitan socio-ecological system (Sahraoui et al. 2021). For a color version of this figure, see www.iste.co.uk/salles/estuarinecities.zip*

The stakeholders identified form nine categories:

– land stakeholders;

– stakeholders of the regulatory framework (e.g. the deconcentrated services of the State, Bordeaux Métropole);

– expertise and knowledge stakeholders (e.g. researchers, naturalist associations, biodiversity observatories, urban planning agencies);

– landowners (e.g. foresters, wine growers, farmers);

– civil society and users (e.g. residents, tourists, citizen associations and groups, neighborhood committees);

– businesses;

– public and private developers (e.g. public works companies);

– elected officials;

– institutional managers (e.g. public forest managers).

The relationships linking stakeholders and their interactions with resources show, for example, how elected officials interact with land use and occupation by defining their status and trajectory. Contrary to a classical ARDI approach, here resources can also interact with stakeholders, through intangible resources and a regulatory framework imposed by public action instruments.

This work of making visible the socio-ecological interdependencies and comparing the visions of each category of stakeholders was a prerequisite for the development of prospective scenarios for the evolution of the territory.

9.3. Scenarios of metropolitan trajectories: contrasted political–ecological footprints

Based on the shared cognitive model of the socio-ecological functioning of the Bordeaux Métropole territory, which emerged from the first workshop, scenarios were developed during a second workshop. The collective construction of scenarios is a means of integrating a multiplicity of ideas and thoughts into holistic images, giving meaning to possible futures (Rasmussen 2008). It does not aim to predict the future, but rather to estimate the possible impacts of various urban development trajectories, considering the multiplicity of stakeholder viewpoints involved in the process (Etienne et al. 2011). The scenarios are based on a graphic representation of urban development trajectories, accompanied by narratives and maps of changes in the territory. This method of formalizing trajectories is inspired by environmental prospective works (*Futuribles*, Mermet 2005). The method of building prospective scenarios has been tested in the context of a collaborative research agenda (de Godoy Leski et al. 2018) or integrative research projects (Labbouz et al. 2017). This participatory approach provides participants with a cognitive and normative framework distanced from the norms and practices of their professional experiences, and allows for analysis of "the ability to take into account simultaneously a large number of complex interdependencies of an accounting, institutional, structural, and logical nature" (Charpin 1983, p. 103, *author's translation*).

Traditionally, the scenario development phase was based on observation and documentation of current trends, weak signals and known major urban development projects. The objective was to collectively imagine various modes of urban governance, reconciling the conservation of biodiversity and the urban development of the Bordeaux Métropole. The time horizon was collectively determined to be 15–20 years, a long enough period to visualize environmental and urban changes in concrete terms, while offering the possibility of tangible development and planning choices. Within the framework of the prospective exercise, the projection into the short and medium term (around 2035) allows stakeholders to cognitively grasp the city in the making.

This co-constructed work was framed by a matrix proposing two axes of urbanization trajectory: (1) according to spatial dimensions and (2) according to antagonistic modes of governance (Figure 9.2):

1) in a spatial dimension, *metropolization*, which designates a concentration of human, technical and natural resources (de Godoy Leski 2021), is opposed to *metapolization*, defined as the capacity of cities to maintain their ordinary function through action on their margins (Ascher 1995);

2) regarding modes of governance, *adaptive governance* (Folke et al. 2005; Folke 2007), defining short-term adjustment strategies implemented in the face of uncertain environmental crises, is opposed to *anticipatory governance* (Boyd et al. 2015), which considers in a longer-term political vision several strategies and decisions in a range of possible futures.

At the intersection of the axes of the matrix, the participants were asked to position three components relating to (1) the urban development (structures and dynamics of urban spaces), (2) ecological and landscape configurations (structures and dynamics of natural and semi-natural habitats) and (3) governance through tools for public action. These components are positioned by the participants according to their level of importance, according to whether they are structuring (first level of importance), supporting (second level of importance) or contingent (third level of importance), and according to an order that may vary according to the scenarios.

Along with the graphic medium, participants were asked to narrate the scenarios with the goal of memorizing and communicating alternative futures in a convincing manner (Schwartz and Ogilvy 1998; Rasmussen 2008). Through these narratives, incorporating key variables and involving spatial and temporal changes for the territories, participants described the end state of the scenarios, as well as the processes or changes leading, component by component, to that end state, and according to the defined trajectory (de Godoy Leski et al. 2018). From this perspective, the narratives function as a cognitive tool for decision-making based on anticipating the ecological impacts of urbanization.

Participants were instructed to successively construct:

1) a realistic trend scenario (Iwaniec et al. 2020), designed to explore territorial dynamics as they might occur without a change in governance;

2) several dystopian and utopian scenarios (Hjerpe and Linnér 2009), intended to explore negative and positive dynamics;

3) an ideal–realistic desirable scenario (Wiek and Iwaniec 2014; Iwaniec et al. 2020), in reaction to the previous scenarios and intended to express the results of a resilient transformation of the territory.

In total, five scenarios were co-constructed (Figure 9.2): a trend scenario, two dystopian scenarios, a utopian scenario and a transformative scenario in response to the previous four. The trajectories show a utopian projection that opposes existing metropolitan dynamics. The desirable scenario, intermediate to the utopian visions, and the trend scenario are also in opposition to the dystopian scenarios.

These trajectories and associated narratives have been complemented by participatory mapping. These maps make it possible to project and spatialize the changes in the territory in the form of changes in the state of land use that each scenario implies of the city (Figure 9.3). In particular, they highlight the differentiated footprints of urban development in the city in the case of the trend and dystopian scenarios (more or less dense or spread out), and the effects of ecological restoration at the urban–rural interface in the case of the utopian scenario.

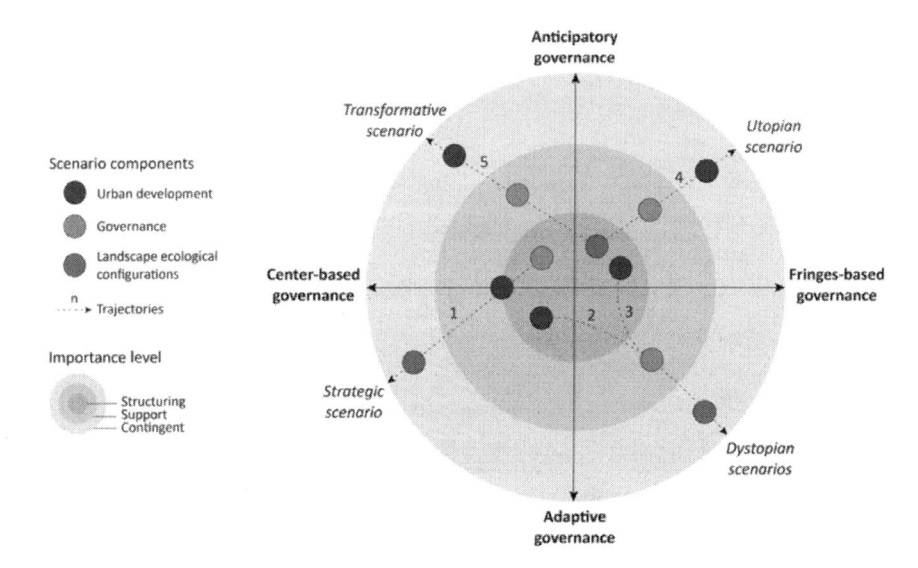

Figure 9.2. *Scenario trajectories co-constructed by the participants (Sahraoui et al. 2021). The narratives of each of the scenarios (see below) are presented in the order of the components (from structuring to contingent), according to the hierarchies established by the participants, independently, and for each of them. For a color version of this figure, see www.iste.co.uk/salles/estuarinecities.zip*

9.3.1. *Strategic scenario: "Dense City" or the Return of the Rhine Model (Scenario 1)*

The metropolitan and anticipatory governance is the structuring component of the trend scenario. It is embodied by Bordeaux Métropole, which is asserting itself as a regional, national and international city. This governance is based on the implementation of public policies aimed at increasing the population to one million inhabitants, within the restricted space of the metropolitan urban envelope.

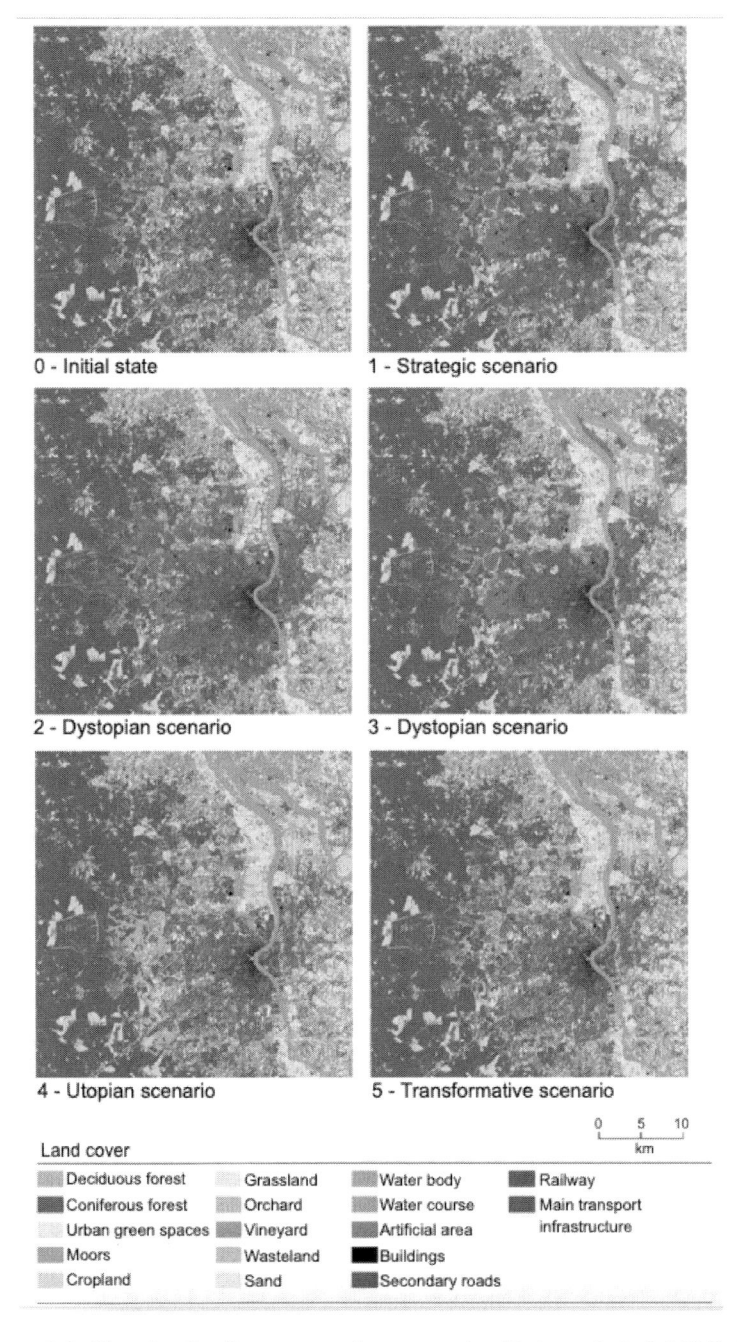

Figure 9.3. *Mapping the five prospective scenarios (Sahraoui et al. 2021). For a color version of this figure, see www.iste.co.uk/salles/estuarinecities.zip*

This proactive governance, driven by anticipation, is based on the programming of major territorial and urban projects along the river, aimed at densifying the hypercenter with the urban fringes of the city (OIM Innocampus, OIM Aéroparc, PPEANP des Jalles).

In terms of urban development, a restriction area established by the SCoT (a territorial coherence instrument), which sets the limits for urbanization, is respected. At the fringe of this planning, a more spontaneous urban growth is taking place, on the periphery (right bank and left bank), guided by the dynamics of real estate development according to the needs of housing and guided by the economy. Urban densification is accompanied by strong social expectations in terms of the preservation of urban landscapes. The PLUi inscribes the obligation to respect a proportion of 50% urbanized spaces and 50% natural and agricultural spaces, all the U and AU zones are urbanized, while the A and N spaces, located on the margins, are viewed as restricted areas. The issue on the development of industrial wastelands puts pressure on ecological development strategies.

The ecological and landscape configurations, and in particular the ecological continuities (green and blue infrastructures) translated in the SCoT and the PLUi, constitute adjustment variables for development operations, and are only treated in urban projects if there are strong regulatory constraints in the PLUi. This consideration of the ecological dimension becomes a variable depending on the restriction area. The legal pressure exerted by the joint action of the urban planning and environmental rules frames large projects in their consideration of ecological and landscape configurations, while small project zones or spontaneous urbanism are exempted. Ecological continuities are perceived as important only for their aesthetic character, and as components of the living environment and the well-being that they provide to the inhabitants. This utilitarian vision of nature is mainly put into space within the urban fringes, while the heart of the city continues to be shielded from the continuity of existing developments. Faced with urban densification, the peripheral natural spaces, as well as the squares, parks and urban gardens are subject to strong anthropic pressure linked to the intensification of their uses. The over-frequentation of these spaces, the result of their attractiveness for leisure activities, is accompanied by environmental degradation through pollution, waste and the trampling of plants.

9.3.2. *Dystopian scenarios: "city–nature opposition" (Scenario 2) and "city–nature interweaving" (Scenario 3)*

Urban development, a structuring component of these scenarios, takes the form of a hyper-densification of residential spaces, shops, services and businesses within the city. Urban development is implemented in several ways: through a verticality of the city center accompanied by: (1) urban sprawl within the limits of the city (Scenario 2); and (2) urban sprawl beyond its institutional limits and structured by ordinary territorial mobility (towards the north, Spain, the Arcachon Basin, the Entre-deux-mers) (Scenario 3).

A metapolitan and adaptive governance supports these dynamics and focuses on regulations to respect the dichotomy between urban intra-metropolitan space and natural extra-metropolitan space. The implementation of (ERC-type) compensation taxes for intra-metropolitan projects allows for both the restoration of natural environments in the vicinity, their accessibility and their development (parking lots, etc.) for recreational activities.

The ecological and landscape configurations then take several forms: a peripheral green belt or green corridors in the urban interstices. In the first case (Scenario 2), the preservation and/or management of the natural environments in the vicinity is accompanied by an integral reservation allowing for the ecological functions of these environments to be preserved. In the second case (Scenario 3), no natural space is systematically protected; however, the preservation of green corridors within the city allows for the maintenance of visual landscape amenities.

9.3.3. *Utopian scenarios: "radical ecological restoration" (Scenario 4.1) and "optimal reconciliation" (Scenario 4.2)*

The ecological and landscape configurations constitute the structuring component of the scenario. The preservation and restoration of ecological continuity and the functionality of natural, terrestrial and aquatic habitats are the main drivers. This translates into the maintenance of existing and remaining corridors and habitats, even those without connections. Concerning aquatic environments, the restitution operations result in the depollution of existing rivers, as well as the reopening of the surrounding environments through the restoration of old rivers, by the destruction of storm drains or infrastructures affecting the free circulation of water and the naturalness of the banks. The restoration of the alluvial zones of the Garonne, the estuary, the *Jalles* and the *Eau Bourde* is also valued as part of the image of the city being promoted by Bordeaux Métropole. The strategy of ecological development politically assumed implies the restoration of the banks and the maintenance of water diversion zones composed of meadows and/or recreational areas. Concerning the terrestrial and transitory environments, the restoration of the ecological functions of the forests of the *Landes plateau* requires the maintenance and planting of deciduous trees to counter the dynamics of intensive reforestry (e.g. the strategy of planting maritime pine to dry out the soil starting in the 19th century). The wetlands of the western part of the city (including the network of lagoons) are also preserved and restored.

Governance is a supporting component at specific scales and with specific modes of action. In the first case (Scenario 4.1), this governance relies on stakeholders with a broader metapolitan scope of action: either on the scale of the region with strong competencies to reconcile urban planning and the preservation/restoration of ecological continuities or on the scale of the department functioning as a *metapole*, or a new metapolitan actor yet to be imagined. For strong governance, the establishment of operations of metropolitan interest (OIM in French) intended for ecological planning (as opposed to OIMs for urban planning projects) consolidates this trajectory to define the economic stakes according to the environmental stakes. In a second case (Scenario 4.2), governance is based on original and dedicated forms of participative democracy.

The urban development constitutes here a contingent component of the ecological and landscape configurations. Ecological preservation and restoration require an adaptation of urban forms, either in the direction of a de-densification of the city through the relocation of the inhabitants or in the direction of concentrated urbanization into non-natural spaces. In the first case (Scenario 4.1), the restoration of the riverbanks and alluvial zones is accompanied by the deconstruction of certain residential areas, and thus, a limitation for the growth of the metropolitan population is implemented. This scenario requires a decentralization of economic activities and services (culture, education, health, etc.) outside of Bordeaux, by revitalizing certain areas that are asserting themselves as poles of equilibrium (coastal or estuary cities, Médoc, Entre-deux-mers, nearby the medium-sized cities of Langon, La Réole and Libourne) and other local urban poles (Angoulême, Poitiers, etc.) allowing inhabitants to live and work while limiting their center–periphery travel. The implementation of soft mobility infrastructures and the reclaiming of the river and the estuary (which can serve as a support for mobility in a north–south axis) become indispensable conditions. In the second case (Scenario 4.2), the urban configuration is based on the "city-archipelago" model, with the development of poles closer to the central city. Urban development takes place in the interstitial spaces of ecological continuities, by urbanizing already equipped spaces. These new constructions are inspired by natural habitats and ecological engineering innovations (constructions on stilts, green roofs, etc.).

9.3.4. *Transformative scenario: resilient city (Scenario 5)*

The development of Bordeaux Métropole is guided by an objective to be attractive and for its population to reach one million inhabitants by 2030, thereby increasing its vulnerability to environmental events. However, Bordeaux Métropole considers that this objective requires the ecological and landscape configurations to be included in a green and blue infrastructure extended to the Gironde department and its estuary. A reopening of the urban waterways is established, accompanied by solutions of depollution and sanitation from the existing overhead and underground hydrographic network. The ecological and landscape framework of the city is strengthened through the enhancement of urban ecological niches, while connecting these niches to the functioning of ecosystems on a more macroscopic scale. Biodiversity is understood in a dynamic way, as opposed to the vision of a fixed ecology. Emphasis is placed on the knowledge of natural spaces and their connectivity in order to improve ecological restoration operations. This requires visibility of the ecological functionalities of environments and ecological continuities through fine-scale mapping, which can be integrated into urban planning documents.

The political reflections linked to metropolitan governance extend to the peripheral territories. On the one hand, the narratives of the territory constitute an important tool for political support by elected officials, aiming to reconcile the development of the city with its environmental challenges. On the other hand, the limits of the "project-based" approach and the influence of political power games are recognized. These point to the need for an overall vision of the territory, which requires more cross-cutting and long-term forms of organization and action (cross-cutting institutional services, territorial time agencies, etc.). The political leadership on ecological issues is assumed by Bordeaux Métropole and its peripheral territories, which recognize the impasse of public environmental concern, previously driven by an ecosystem services approach.

The establishment of a tax system encourages municipalities to protect natural habitats and species. The maintenance of the ecological framework is conditioned by a regulatory framework along with transparency where it concerns compensation sites. Offset managers integrate concerted/shared and open governance, which adjusts according to the ecological risks. In parallel, environmental health is set up as a guiding principle for the construction of territories in terms of quality of life.

The temporality of development and construction projects, coupled with the ecological, architectural and legal constraints that frame the logic of planning, does not allow for the objective of population growth to be achieved. Urban development is now subject to ecological and landscape configurations, and to the injunctions of new governance. On the contrary, the coherence of urban project sites is achieved through the use of a tool intended for project developers, in order to stabilize an agreement between municipalities with initially divergent interests. The availability of land is put under pressure by the central city (Bordeaux) which is becoming denser. The transport infrastructures, and in particular the tramway and soft mobility routes, are seen as vectors of new ecological functions.

9.4. Conclusion

This chapter proposed an analysis on the governance of urban biodiversity by means of an original approach through co-construction with stakeholders within the territory, and evolution scenarios for the estuarine city. This approach, based on narratives accompanied by (carto)graphic modeling, has made it possible to highlight the articulation between elements relating to the urban fabric, governance and the ecological and landscape configurations of the city. In particular, it has shown the need for anticipatory governance of the challenges and dynamics of biodiversity. Indeed, the trend and dystopian scenarios, defined by the participants, are linked to the action principles of adaptive governance, while the utopian and desirable scenarios are linked to the action principles of anticipatory governance. The co-constructed narratives reveal, in particular, the current impasses of a public action that only integrates biodiversity into urban development projects a posteriori, through mitigation or compensation measures characteristic of adaptive policies.

In contrast to these trends, the participants expressed a need for anticipation, which translates into the construction of desirable and utopian future alternatives based on the primacy of ecological and landscape configurations, through stronger consideration of biodiversity conservation issues in metropolitan governance. During this process, the stakeholders expressed the need for a better understanding and stressed the importance of more integrated and collaborative action to address these cross-sectoral issues. They also expressed their willingness to advocate for the desirable scenario, which may involve questioning some of the planned urban developments. By breaking *path dependency* with the support of strong governance, a paradigm shift could, in fact, consist of moving from a vision of the a posteriori integration of biodiversity in development to a vision of biodiversity as a primary dynamic process that determines urban development choices. This projection exercise, in addition to having made it possible to reveal different options, showed how cognitive work on the issues of urban biodiversity preservation redraws a Bordeaux Métropole inserted in a network of connections and extensive socio-ecological interdependencies.

9.5. References

Ascher, F. (1995). *Métapolis ou l'avenir des villes*. Odile Jacob, Paris.

Bordeaux Métropole (2011). 5 sens pour un Bordeaux métropolitain. Report, Bordeaux.

Bordeaux Métropole (2013a). 50 000 logements autour du transport collectif. Report, Bordeaux.

Bordeaux Métropole (2013b). 55 000 hectares pour la nature. Report, Bordeaux.

Bordeaux Métropole (2016). Espèces de Métropole. Atlas de la biodiversité. Report, Bordeaux.

Boyd, E., Nykvist, B., Borgström, S., Stacewicz, I. (2015). Anticipatory governance for social-ecological resilience. *Ambio*, 44(1), 149–161.

Charpin, J.-M. (1983). Les projections macroéconomiques quantifiées. *Futuribles*, 70.

Étienne, M. (2010). *La modélisation d'accompagnement. Une démarche participative en appui au développement durable*. Quae, Versailles.

Étienne, M., Du Toit, D., Pollard, S. (2011). ARDI: A co-construction method for participatory modeling in natural resources management. *Ecology and Society*, 16(1) [Online]. Available at: http://www.ecologyandsociety.org/vol16/iss1/art44/.

Folke, C. (2007). Socio-ecological system and adaptive governance of the commons. *Ecological Research*, 22(1), 14–15.

Folke, C., Hahn, T., Olsson, P., Norberg, J. (2005). Adaptive governance of social-ecological systems. *Annual Review of Environment and Resources*, 30, 441–473.

de Godoy Leski, C. (2021). Vers une gouvernance anticipative des changements globaux. L'emprise des interdépendances socio-écologiques sur une métropole estuarienne. Bordeaux Métropole et l'estuaire de la Gironde. PhD Thesis, Université de Bordeaux, Bordeaux.

de Godoy Leski, C., Marquet, V., Salles, D. (2018). Sociologie et recherche inclusive : prospective collaborative pour un agenda de recherche sur l'eau. *Sociologies pratiques*, 37(2), 25–38.

Hjerpe, M. and Linner, B.O. (2009). Utopian and dystopian thought in climate change science and policy. *Futures*, 41(4), 234–245.

Iwaniec, D.M., Cook, E., Davidson, M.J., Berbes-Blazquez, M., Georgescu, M., Krayenhoff, E.S., Middel, A., Sampson, D.A., Grimm, N.B. (2020). The co-production of sustainable future scenarios. *Landscape and Urban Planning*, 197, 103744 [Online]. Available at: https://doi.org/10.1016/j.landurbplan.2020.103744.

Jollivet, M. and Pena-Vega, A. (2002). Relier les connaissances, transversalité, interdisciplinarité. *Nature, sciences et sociétés*, 10(1), 78–95.

Labbouz, B., Salles, D., Valette, P. (2017). Les territoires garonnais face aux changements globaux : quatre adaptations possibles en 2050. *Sud-Ouest européen*, 44, 71–82.

Lascoumes, P. (2012). *Action publique et environnement*. PUF, Paris.

Leroy, P. (2001). La sociologie de l'environnement en Europe : évolution, champ d'action ambivalences. *Nature, sciences et sociétés*, 9(1), 29–39.

Lussault, M. (2017). *Hyper-lieux*. Le Seuil, Paris.

Mermet, L. (2005). *Étudier des écologies futures : un chantier ouvert pour les recherches prospectives environnementales*. PIE-Peter Lang/EcoPolis, Brussels.

Pinson, G., Reiffers, E., Hirschberger, S. (2018). L'appel à projets urbains "50 000 logements" à Bordeaux : la mise en échec de la métropole stratège. *Métropolitiques*, 1–6 [Online]. Available at: https://www.metropolitiques.eu/L-appel-a-projets-urbains-50-000-logements-a-Bordeaux-la-mise-en-echec-de-la.html.

Rassmussen, L.B. (2008). The narrative aspect of scenario building – How story telling may give people a memory of the future. *Ai & Society*, 19, 229–249.

Sahraoui, Y., de Godoy Leski, C., Benot, M.-L., Revers, F., Salles, D., Van-Halder, I., Barneix, M., Carassou, L. (2021). Integrating ecological networks modelling in a participatory approach for assessing impacts of planning scenarios on landscape connectivity. *Landscape and Urban Planning*, 209, 104039.

Salles, D. (2006). *Les défis de l'environnement. Démocratie et efficacité*. Syllepse, Paris.

Schewenius, M., McPhearson, T., Elmqist, T. (2014). Opportunities for increasing resilience and sustainability of urban social–ecological systems: Insights from the URBES and the cities and biodiversity outlook projects. *Ambio*, 43(4), 434–444.

Schwartz, P. and Ogilvy, J.A. (1998). Plotting your scenarios. In *Learning from the Future*, Fahey, L. and Randall, R. (eds). Wiley, New York.

Touchard, O. (2019). L'action urbaine écologique de Bordeaux Métropole : le plafond de verre de la nature ou la conflictualité tacite des pratiques d'aménagement. PhD Thesis, Université Michel de Montaigne, Pessac.

Wiek, A. and Iwaniec, D. (2014). Quality criteria for visions and visioning in sustainability science. *Sustainability Science*, 9(4), 497–512.

Conclusion

Climate change, the collapse of biodiversity or the Covid-19 health crisis describes the story of the Anthropocene. It is a story about the vulnerabilities to the ever-tightening network of interdependencies between human societies across the planet. At a time when ecological and health crises are questioning the habitability of the Earth (Stengers 2009; Latour 2015; Audier 2020), a new reading of complexity (Morin 2014) has become a necessity in order to respond to the challenges posed by global change. The contemporary ecological crisis reveals more than ever the interconnectedness of networks between all living and non-living things. This ecology of relations (Descola 2019) promotes a vision of the world that prioritizes the observation of flows, constrained or chosen reciprocities, networks without an organizing consciousness, neither a sense of belonging, nor a visible boundary (Grosseti and Barthe 2008).

For the manufacture of cities, the revelation on the generalized interdependencies of environmental problems challenges the classic forms of sectoral expertise and territorial governance. Faced with global changes, metropolises are challenged to imagine adaptation trajectories that extend to their entire territory of influence.

While consideration of interdependencies (Salles 2006; Carter 2018; Offner 2020) is not new, it is at the heart of social–ecological transition approaches and projects (Laurent 2018). These socio-ecological

---

Conclusion written by Denis SALLES, Glenn MAINGUY and Charles DE GODOY LESKI.

interdependencies (de Godoy Leski 2021) invite us to consider estuarine cities and their countryside, as a more or less dense and stretched network of unstable relationships, woven between ecology and society, forming socio-ecological configurations at different spatial and temporal scales. This complex network is composed of several dimensions – ecological, cognitive, institutional, instrumental, spatial and temporal – which contribute to the visibility of a world structured less by borders than by flows. The various case studies developed in this book have shown the need to make these socio-ecological interdependencies visible: between the water for cities and the water for fields, between surface water and groundwater, risk management of the river banks upstream and downstream of the estuary, and between the successive historical periods of implementing activities.

This work of making the socio-ecological interdependencies visible and the urgency of defining credible time horizons so as to preserve the Earth's habitability leads us to reason in the long term. What kind of future do we want? Climate change invites us to renew the dated reference system of the sustainable city in order to open up the capacity to anticipate global changes. Promoting a new and expanded city, "facing towards the future", aims to anticipate its development, biodiversity, quality of life and the safety of its "gray and green" infrastructures.

For estuarine cities, these forms of "adaptive" and "anticipatory" governance (Quay 2010; Rocle et al. 2020; de Godoy Leski 2021; Salles 2022) require innovative and novel approaches that integratively address the making of the city and its adaptation to global change. Anti-participatory governance is presented as a flexible decision-making framework built collectively and in a concerted and inclusive manner. It is based on the exploration of scenarios and pathways of futures[1] that integrate uncertainties (Quay 2010), and the territorial geopolitics between the global regulations of globalization and those at the local level of territory living.

The development of scenario-based approaches invites us to open up the universe of possibilities, from the most conventional options (adaptation by adjustment) to the exploration of alternative paths (adaptation by transformation) (Basset and Fogelman 2013). In order to become real operators of the political agenda, these scenario-based approaches intend to

---

1 https://futureearth.org/initiatives/earth-targets-initiatives/science-based-pathways/.

mobilize, in an integrated way, the objectification tools for modeling and digital simulation, as well as socio-anthropological approaches capable of capturing the subjective dimensions of meanings and emotions. Through these scenario approaches, it is also a question of bringing on board pluralistic social collectives, in order to generate visions of societal transformation in which the ecological, social, economic, cultural and political dimensions are interconnected (Labbouz et al. 2021[2]; Rocle et al. 2020; Rulleau et al. 2021).

How can transformative visions emerge in the face of real-world constraints? How can aspects of knowledge, values, aspirations, conflicting interests, power struggles and also emotions that frame and weave societal dynamics be interwoven into political project? What are the potentials and limits of scenario approaches, and what are the tools and services to deploy them in terms of transitions and the engagement of projects at the scales of sectors, territories, communities and individuals? For the societies of the 21st century, the anti-emergence of climate change is now seen as a step to be taken collectively. The socio-ecological transition is the pathway, knowledge is the compass and the political narrative expresses its sensemaking.

References

Audier, S. (2020). *La cité écologique. Pour éco-républicanisme*. La Découverte, Paris.

Basset, T. and Fogelman, C. (2013). Déjà Vu or Something New? The adaptation concept in the climate change literature. *Geoforum*, 48, 42–53.

Blondel, J. (2008). Changements globaux. *Forêts méditerranéennes*, 29(2), 119–126.

Carter, C. (2018). *The Politics of Aquaculture Sustainability Interdependence, Territory and Regulation in Fish Farming*. Routledge, London.

Charbonnier, P. (2021). *Abondance et liberté. Une histoire environnementale des idées politiques*. La Découverte, Paris.

Cocula, A.-M. and Audinet, E. (2018). *L'estuaire de la Gironde, une histoire au long cours*. Confluences, Bordeaux.

---

2 http://www.adapteau.fr/.

Deldreve, V., Candau, J., Noûs, C. (2021). *Effort environnemental et équité. Les politiques publiques de l'eau et de la biodiversité en France*. Peter Lang, Brussels.

Descola, P. (2019). *Une écologie des relations*. CNRS, Paris.

Driessen, P., Leroy, P., Van Vierssen, W. (2010). *From Climate Change to Social Change. Perspective on Science-Policy Interactions*. International Books, Utrecht.

de Godoy Leski, C. (2021). Vers une gouvernance anticipative des changements globaux. L'emprise des interdépendances socioécologiques sur une métropole estuarienne. Le cas de Bordeaux et l'estuaire de la Gironde. Sociology PhD Thesis, INRAE/Université de Bordeaux, Bordeaux.

Grosseti, M. and Barthe, J.-F. (2008). Dynamique des réseaux interpersonnels et des organisations dans les créations d'entreprises. *Revue française de sociologie*, 49, 585–612.

Guilluy, C. (2014). *La France périphérique. Comment on a sacrifié les classes populaires*. Flammarion, Paris.

Labbouz, B., Lumbroso, S., Vial, I. (2021). Les ressources de la prospective au service de la biodiversité : comment mobiliser les futurs pour les politiques publiques de biodiversité ? Guide, Comprendre pour agir, Office français de la biodiversité.

Latour, B. (2015). *Face à Gaïa. Huit conférences sur le nouveau régime climatique*. La Découverte, Paris.

Laurent, E. (2018). La transition sociale-écologique : récit, institutions et politiques publiques. *Cités*, 4(76), 31–40.

Löw, M. (2015). *Sociologie de l'espace*. Maison des Sciences de l'Homme, Paris.

Marquet, V. (2014). Les voies émergentes de l'adaptation au changement climatique dans la gestion de l'eau. Mises en visibilités et espaces de définition. PhD Thesis, Irstea/Université de Bordeaux, Bordeaux.

Mele, P. (2013). *Conflits de proximité et dynamiques urbaines*. PUR, Rennes.

Mermet, L. and Salles, D. (2015). *Environnement : la concertation apprivoisée, contestée, dépassée ?* De Boeck Supérieur, Louvain-la-Neuve.

Morin, E. (2014). *Introduction à la pensée complexe*. Le Seuil, Paris.

Offner, J.-M. (2020). *Anachronismes urbains*. Presses Science Po, Paris.

Quay, R. (2010). Anticipatory governance. A tool for climate change adaptation. *Journal for Planning Association*, 76(4), 496–511.

Rocle, N., Rey-Valette, H., Bertrand, F., Becu, N., Long, N., Bazart, C., Lautredou-Audouy, N. (2020). Paving the way to coastal adaptation pathways: An interdisciplinary approach based on territorial archetypes. *Environmental Science & Policy*, 110, 34–45.

Rulleau, B., Salles, D., Gilbert, D., le Gat, L.Y., Renaud, E., Bernard, P., Stricker, A.E. (2020). Crafting futures together: Scenarios for water infrastructure asset management in a context of global change. *Water Supply*, 20(8), 3052–3067.

Salles, D. (2006). *Les défis de l'environnement, démocratie et efficacité*. Syllepse, Paris.

Salles, D. (2011). Responsibility based environmental governance. *S.A.P.I.EN.S*, 4.1 [Online]. Available at: http://journals.openedition.org/sapiens/1092.

Salles, D. (2022). Repenser l'eau à l'ère du changement climatique. *Responsabilité et environnement*, 106 [Online]. Available at: http://www.annales.org/re/2022/resumes/avril/08-re-resum-FR-AN-avril-2022.html#08FR.

Salles, D. and Le Treut, H. (2017). Comment la région Nouvelle Aquitaine anticipe le changement climatique ? *Sciences, eaux & territoires*, 22, 14–17.

Sire, P. (2016). *Le fleuve impassible*. Le Festin, Bordeaux.

Stengers, I. (2009). *Au temps des catastrophes. Résister à la barbarie qui vient*. La Découverte, Paris.

Subra, P. (2018). *Géopolitique de l'aménagement du territoire*. Armand Colin, Paris.

List of Authors

Gilles BILLEN
METIS, CNRS
Sorbonne Université
Paris
France

Henrique CABRAL
INRAE
EABX
Bordeaux
France

Cécile CAPDERREY
BRGM
Orléans
France

Jeanne DACHARY-BERNARD
INRAE
ETTIS
Bordeaux
France

Alain DUPUY
EPOC
ENSEGID
Bordeaux INP
France

Michael ELLIOTT
University of Hull
UK

Josette GARNIER
METIS, CNRS
Sorbonne Université
Paris
France

Charles DE GODOY LESKI
LGP, CNRS
Université Paris 1
Panthéon-Sorbonne
UPEC
Thiais
and
CED
Université de Bordeaux
France

Julia LE NOË
Géologie de l'ENS
ENS
Paris
France

Mario LEPAGE
INRAE
EABX
Bordeaux
France

Glenn MAINGUY
CED
Université de Bordeaux
France

Thierry OBLET
CED
Université de Bordeaux
France

Yohan SAHRAOUI
ThéMA
Université Bourgogne
Franche-Comté
Besançon
France

Denis SALLES
INRAE
ETTIS
Bordeaux
France

Nabil TOUILI
Université Paris-Saclay
France

Jean-Paul VANDERLINDEN
Université Paris-Saclay
France
and
University of Bergen
Norway

Florian VERGNEAU
École d'économie de Toulouse
France

Aude VINCENT
University of Iceland
Reykjavik
Iceland

Index

A, B

adaptation, 45
allegorical value, 115
amenities, 46
    green, 47
anthropization, 136
anthropogenic pressures, 73
aquifer, 15
artificialization, 169
attenuation, 33
BiodiverCité, 176
biodiversity, 14, 69
Bordeaux, 117
    Métropole, 23

C, D

catchment area, 4
change(s), 18
    climate, 45
    global, 46
cities, 191
    estuarine, 45
cohesion, 124
Conférence Permanente Loire, 147
connection, 93
controversy, 5
countryside, 121
degradation, 69

deontology, 98
development, 46
dimension
    estuarine, 137
    symbolic, 116
drainage ditches, 31

E, F

eco-engineering, 71
ecological territory, 149
ecology, 191
    territorial, 155
ecosystem(s), 29, 70
    services, 70
environmental mutations, 93
environmentally friendly engineering, 30
environments, 69
Estuarium association, 139
estuary, 155
    Gironde, 82
    Loire, 135
    Mondego, 83
ethical theories, 97
ethics
    deliberation, 103
    of nature, 105
    virtue, 102

exposure to risk, 47
flood zone (*see also* Nantes *and* risk), 61
frame of reference, 96
French society, 115

G, H

GIP Loire Estuaire, 136
governance, 45
   territorial, 93
government, 115
Great Citizen Debate, 145
habitats, 69

I, L

Île de Nantes, 141
impacts, 69
instruments, 94
inter-terrioriality, 130
interdependence of territories, 116
interface, 156
intertidal mudflats, 69
Landes du Médoc, 28

M, N

meta-narratives, 98
metabolism, 169
metapole, 190
Mission Loire, 148
modeling, 27
Nantes (*see also* Île de Nantes), 143
   floods, 142
narrative, 96
   historical, 115
natural disaster, 62
nature-based solutions, 71

P, R

participatory tool, 145
periphery, 93

port, 167
prospective, 176
protection, 70
residential choices, 46
resources, 180
   aquatic, 69
restoration, 70
risk
   natural, 142
   flooding, 45
river-port cities, 143

S, T, U

scenarios, 183
Scheldt, 85
Seine, 155
sense, 96
socio-ecosystems, 155
solidarity, 115
storm Xynthia, 94
submersion/submergence, 46, 156
sustainable urban development, 45
territorial equality, 116
territorialized approaches, 94
trajectories, 176
urban fabric, 192

V, W

vulnerability, 45, 191
watershed, 155
ways of judging, 93

Other titles from

iSTE

in

Ecological Science

2023

GAID Kader

Drinking Water Treatment 1: Water Quality and Clarification

Drinking Water Treatment 2: Chemical and Physical Elimination of Organic Substances and Particles

Drinking Water Treatment 3: Organic and Mineral Micropollutants

Drinking Water Treatment 4: Membranes Applied to Drinking Water and Desalination

Drinking Water Treatment 5: Calco-carbonic Equilibrium and Disinfection

2022

AMIARD Jean-Claude

Management of Radioactive Waste (Radioactive Risk Set – Volume 5)
Marine Radioecology (Radioactive Risk Set – Volume 6)

GODET Laurent, DUFOUR Simon, ROLLET Anne-Julia

The Baseline Concept in Biodiversity Conservation: Being Nostalgic or Not in the Anthropocene Era

GRISON Claude, CASES Lucie, LE MOIGNE Mailys,
HOSSAERT-MCKEY Martine
Photovoltaism, Agriculture and Ecology From Agrivoltaism to Ecovoltaism

LÉVÊQUE Christian
Biodiversity Erosion: Issues and Questions

2021

FLEURENCE Joël
Microalgae: From Future Food to Cellular Factory

AMIARD Jean-Claude
*Disarmament and Decommissioning in the Nuclear Domain
(Radioactive Risk Set – Volume 4)*

JOLY Fernand, BOURRIÉ Guilhem
*Mankind and Deserts 1: Deserts, Aridity, Exploration and Conquests
Mankind and Deserts 2: Water and Salts
Mankind and Deserts 3: Wind in Deserts and Civilizations*

MUTTIN Frédéric, THOMAS Hélène
Marine Environmental Quality

PIOCH Sylvain, SOUCHE Jean-Claude
*Eco-design of Marine Infrastructures: Towards Ecologically-informed
Coastal and Ocean Development*

2020

BRUSLÉ Jacques, QUIGNARD Jean-Pierre
*Fish Behavior 1: Eco-ethology
Fish Behavior 2: Ethophysiology*

LE FLOCH Stéphane
Remote Detection and Maritime Pollution: Chemical Spill Studies

MIGON Christophe, NIVAL Paul, SCIANDRA Antoine
The Mediterranean Sea in the Era of Global Change 1: 30 Years of Multidisciplinary Study of the Ligurian Sea
The Mediterranean Sea in the Era of Global Change 2: 30 Years of Multidisciplinary Study of the Ligurian Sea

ROSSIGNOL Jean-Yves
Climatic Impact of Activities: Methodological Guide for Analysis and Action

2019

AMIARD Jean-Claude
Industrial and Medical Nuclear Accidents: Environmental, Ecological, Health and Socio-economic Consequences
(Radioactive Risk SET – Volume 2)
Nuclear Accidents: Prevention and Management of an Accidental Crisis
(Radioactive Risk SET – Volume 3)

BOULEAU Gabrielle
Politicization of Ecological Issues: From Environmental Forms to Environmental Motives

DAVID Valérie
Statistics in Environmental Sciences

GIRAULT Yves
UNESCO Global Geoparks: Tension Between Territorial Development and Heritage Enhancement

KARA Mohamed Hichem, QUIGNARD Jean-Pierre
Fishes in Lagoons and Estuaries in the Mediterranean 2: Sedentary Fish
Fishes in Lagoons and Estuaries in the Mediterranean 3A: Migratory Fish
Fishes in Lagoons and Estuaries in the Mediterranean 3B: Migratory Fish

OUVRARD Benjamin, STENGER Anne
Incentives and Environmental Policies: From Theory to Empirical Novelties

2018

AMIARD Jean-Claude
Military Nuclear Accidents: Environmental, Ecological, Health and Socio-economic Consequences
(Radioactive Risk SET – Volume 1)

FLIPO Fabrice
The Coming Authoritarian Ecology

GUILLOUX Bleuenn
Marine Genetic Resources, R&D and the Law 1: Complex Objects of Use

KARA Mohamed Hichem, QUIGNARD Jean-Pierre
Fishes in Lagoons and Estuaries in the Mediterranean 1: Diversity, Bioecology and Exploitation

2016

BAGNÈRES Anne-Geneviève, HOSSAERT-MCKEY Martine
Chemical Ecology

2014

DE LARMINAT Philippe
Climate Change: Identification and Projections